Computational Molecular Science and Engineering Forum 2015

Core Programming Area at the 2015 AIChE Annual Meeting

Salt Lake City, Utah, USA
8-13 November 2015

ISBN: 978-1-5108-1855-2

Printed from e-media with permission by:

Curran Associates, Inc.
57 Morehouse Lane
Red Hook, NY 12571

Some format issues inherent in the e-media version may also appear in this print version.

Printed by Curran Associates, Inc. (2016)

For permission requests, please contact AIChE
at the address below.

AIChE
120 Wall Street, FL 23
New York, NY 10005-4020

Phone: (800) 242-4363
Fax: (203) 775-5177

www.aiche.org

Additional copies of this publication are available from:

Curran Associates, Inc.
57 Morehouse Lane
Red Hook, NY 12571 USA
Phone: 845-758-0400
Fax: 845-758-2634
Email: curran@proceedings.com
Web: www.proceedings.com

TABLE OF CONTENTS

422389 A CFD Integrated Multiscale Method for Modeling Ethylene Oxidation Fixed Bed Reactors

Abstract:

Ethylene oxide is one of the most important petrochemical intermediates with annual production of more than 14 billion ton. Currently, ethylene oxide is exclusively produced through ethylene oxidation over silver catalyst. The reaction is fast and highly exothermic with carbon dioxide as the main by-product. Under industrial conditions selectivity toward ethylene oxide is 65%-80%. Therefore, an increase in selectivity not only reduces greenhouse gases emissions but also results in more efficient use of existing materials and utilities, and eventually a sustainable design.

Extensive research on the ethylene oxidation mechanism over silver catalyst has been done. It is suggested that oxametallacycle is an important intermediate in the mechanism. Oxametallacycle is isomerized to ethylene oxide and acetaldehyde in two different pathways. Acetaldehyde is then oxidized, and produces CO_2 and H_2O, rapidly.

In the present study, to address selectivity loss in an EO fixed bed, computational fluid dynamics is coupled with detailed kinetics and effective pellet diffusivity. This multiscale approach gives a new insight into this partial oxidation reaction. Initially, a reduced microkinetic model is developed based on a 17 step microkinetic model by Stegelmann et al. (2004). The reduced model is implemented in an ANSYS FLUENT user-defined function to calculate reaction rates and surface sites coverages. 3D CFD simulations are carried out on a simple geometry of 9 spherical solid particles inside a fluid zone. Subsequently, the effects of the flow and transport on elementary steps are studied. Our simulations show that under certain operating conditions adsorbed ethylene oxide favors the reverse reaction, and converts to oxametallacycle, which then contributes to carbon dioxide production. This phenomenon results in selectivity loss inside the catalyst particles. In our subsequent work, this method is extended to larger scale models of randomly packed beds of spheres.
In previous studies, it was assumed that the effective diffusivity of species inside the catalyst particle is constant. A simplification of the Dusty Gas Model derived by Hite and Jackson (1977) was used to obtain a Fickian form to model diffusion inside the solid particles. This simplification required the assumption that species flux ratios are equal to the corresponding stoichiometric ratio. However, this is a questionable assumption in 3D simulations. In this study, to examine the validity of that assumption, an anisotropic diffusivity tensor user-defined function is made and implemented into the CFD method. Hite and Jackson's model is used without the need for any further assumptions, i.e. the flux for each species is calculated directly in each computational cell. It is observed that the reaction rate and species concentration profiles inside the particles deviate from those calculated by the method with the simplifying assumption. These results show the importance of taking into account the details of diffusion limitations inside the solid particles.

This approach bridges between macro and micro scale modeling, and shows that coupling detailed transport and kinetics is unavoidable to develop an accurate model of catalytic reactions, and achieve a superior understanding of these systems.

431709 Multi-Scale CFD+Kmc Simulation of Methanol Partial Oxidation over Ceria in a Fixed Bed

Abstract:

Multi-scale CFD+kMC simulation of methanol partial oxidation over ceria in a fixed bed

Qingang Xiong, Sreekanth Pannala, Stuart C. Daw, Aditya Savara, Steven H. Overbury

Oak Ridge National Laboratory, Oak Ridge, Tennessee 37831, USA

Emails: xiongq@ornl.gov (Q. Xiong), pannalas@gmail.com (S. Pannala), dawcs@ornl.gov (S. C. Daw), savaraa@ornl.gov (A. Savara), overburysh@ornl.gov (S. H. Overbury)

Abstract A research project aiming to achieve multi-scale computational models of catalytic methanol partial oxidation over ceria (CeO2) was initiated for better utilization of clean energy. Relevant reactor design and optimization of operating conditions require fundamental understanding on gas flow and heterogeneous reactions. Computational fluid dynamics (CFD) can complement experiment to provide spatiotemporal details for reactor design and optimization with a reduced developing cycle. However, the underlying reaction mechanisms of methanol partial oxidation on the surface of ceria remain unclear, which hinders an accurate prediction of reactor performance by CFD. In this study, a global reaction pathway which was proposed based on the information from both experiments and literature was employed with CFD to simulate a laboratory-scale fixed bed for isothermal ceria catalyzed methanol partial oxidation. In the corresponding experiment, CeO_2 was diluted with silica sand to form the fixed bed, and 5 % methanol/ Ar was flowed from the reactor bottom. The reactor temperature was varied from 573 K to 673 K to realize six isothermal conditions. The pure macro-scale CFD results were corrected by the detailed information obtained from the micro-scale kinetic Monte Carlo simulation on the ceria surface. A compound wavelet matrix approach was employed to realize the multi-scale coupling. The corrected global reaction kinetics was then employed to simulate product composition and yields under different operating conditions and compared with experiment for the validation of the multi-scale coupling strategy.

Research sponsored by the Laboratory Directed Research and Development Program of Oak Ridge National Laboratory, managed by UT-Battelle, LLC, for the U. S. Department of Energy

429665 Analyzing and Compressing the Parametric Space for Modeling and Design of Heterogeneous Catalytic Systems

Abstract:

Heterogeneous catalytic reactions can involve several gaseous and surface species and numerous interlinking reactions. The development of microkinetic models and identification of optimal catalyst properties requires detailed evaluation of the energetics of the reaction network at the atomic scale and relating it to the macroscopic observables such as rate and selectivity; typical approaches include using direct input or linear correlations developed from density functional theory. A universal property of nonlinear systems is that model response is dominated by a few important (or "stiff") and several unimportant (or "sloppy") parameters [1]. Heterogeneous catalytic systems, too, are expected to be similar.

In this study, we adopt a derivative-based global sensitivity analysis and hierarchical clustering based method to elucidate several heterogeneous catalytic systems to show that: (a) few parameters – binding energy of species and activation barrier of elementary reaction steps – are globally sensitive and form a superset of locally sensitive parameters, (b) the span of parametric sensitivity can vary several orders of magnitude, and (c) several parameters are statistically correlated and hence form clusters of "similar" parameters. This analysis shows that the dimensions of the microkinetic model can be significantly reduced and the catalyst design space compressed prior to applying further simplifying physico-chemical approximations such as semi-empirical correlations of the energetics. The resultant reduced model and compressed parameter space can then be the basis for both mechanistic analysis and catalyst design.

Our proposed method and application to representative heterogeneous catalytic systems will be discussed.

1. Machta, B. B., Chachra, R., Transtrum, M. K., and Sethna, J. *Science* (2013), 342, 6158, 604-607

420032 Up-Scaling of Transport Processes in an Irregular Electrode Pore of a Lithium Ion Battery

Abstract:

Lithium ion batteries used in automobiles industries to produce electric vehicles and hybrid electric vehicles are mostly made of porous electrode composites. The transport processes occurring in the pores of a porous electrode affects battery performance therefore, the performance of batteries can be increase by optimizing the electrode morphology. In most cases the electrode morphology is of irregular structures which are often be simplified to ideal rectangular or cylindrical models. The present work utilized an irregular electrode pore geometry (diverging pore) to understanding the effect of the deviation to the regular pore geometry on Li ion diffusion process in the presence of electric field and electrochemical reaction in the pore of an electrode. The area-averaging upscaling approach of homogenization is used to convert the microscopic two dimensional Nernst-plank equation to the macroscopic level. A microscopic scale diffusion-reaction and electromigration model is derived for a pore domain of both rectangular and cylindrical geometries. Afterwards, the area-averaging approach is applied to this model in order to obtain a macroscopic level pore model in terms of area-averaged concentration and effective transport parameters. By solving this model one can demonstrate the effects of diverging angles on the concentration profile in the pore domain. This information is useful to understand a system-level behavior. The effect of electrochemical reaction on the concentration profile is analyzed for zero order electrochemical reaction. The limiting cases of the diverging geometries developed models are compared with the ideal cases in order to verify the up-scaled diverging geometries pore solutions. The results indicates the importance of potential deviations of the idealized geometry when used to predict system level performance. The findings of this work are important in modeling and simulation of the performance of the electrode-base systems since the current modeling efforts have not assessed these potential deviations. These deviations could lead to important system prediction errors.

431087 Molecular Modeling of Pyrolysis Induced Shrinkage of Cellulose

Abstract:

Pyrolysis of lingocellulosic biomass could become a significant source of renewable chemicals and fuels in the near future. In addition to bio-oil production, fast pyrolysis of biomass produces significant quantities of synthesis gases and char. The processes that control these events are complex and not yet well understood. Among many unanswered questions is that regarding the mechanism(s) responsible for pyrolysis induced shrinkage? Ultimately, shrinkage affects transport properties including porosity and thermal conductivity, and is in-part responsible for the morphology and properties of the resulting biochar, a valuable by-product. The purpose of this study is to use molecular modeling to simulate the effect of shrinkage using model cellulose structures and recent knowledge regarding the mechanism of cellulose thermal degradation. Materials Studio was used to create and optimize a model for crystalline and amorphous cellulose. To simulate the effect of shrinkage, cellulose models were optimized first and then *molecular* fragments were removed from the cellulose matrix as test bed. Using the FORCITE module and the molecular mechanics force field, pcff, molecular dynamics was performed on this cellulosic system at 300 K using an NPT ensemble. The modeled parameters, density, energy and hydrogen bonding at each time step were measured and compared to the initial cellulose structures before and after from which molecular-scale morphological changes were inferred.

421854 Modeling of Diffusion and Catalytic Reactions of Gases in Highly Porous Nanolayers with Dsmc and Openfoam

Abstract:

Modeling of Diffusion and Catalytic Reactions of Gases in Highly Porous Nanolayers with DSMC and OpenFOAM

G. R. Pesch[†,‡], *N. Riefler*[†], *U. Fritsching*[†], *L. Colombi Ciacchi*[§,‡], *L. Mädler*[†]

† -- Foundation Institute of Materials Science (IWT), Department of Production Engineering, University of Bremen, Germany

‡ -- Chemical Engineering -- Recovery and Recycling (VdW), Department of Production Engineering and Center for Environmental Research and Sustainable Technology (UFT), University of Bremen, Germany

§ -- Hybrid Materials Interfaces Group (HMI), Department of Production Engineering and Bremen Center for Computational Materials Science (BCCMS), University of Bremen, Germany

Gas diffusion in at Knudsen numbers (Kn) above 0.1 requires a mathematical treatment based on the Boltzmann equation. Accepted solutions for the description of diffusion in highly porous geometries are for instance the extended Fickian Model or the Dusty-Gas-Model (DGM). Here collisions between molecules are, due to their vast appearance, only statistically considered. Both models require mean geometry values, such as porosity and tortuosity, to describe the porous structure. This results in a simplification of the exact layer geometry, which hinders an effective and accurate description of diffusion processes inside real inhomogeneous 3 dimensional porous layers, such as gas sensors films or catalysts.

In gas sensors for example, a probe gas diffuses into the porous layer. Due to chemical reactions at the surface (rate of reaction is proportional to probe gas concentration) the porous layer resistance changes, which can be measured by two electrodes and back calculated to the probe gas composition. Fundamental for design and response optimization of sensors of this kind is precise knowledge about the amount and exact position of the reaction fronts within the layer.

The Direct Simulation Monte Carlo (DSMC) method is suitable for the simulation of gas diffusion in such sensors as it uses the exact porous geometry to describe diffusion processes by modeling the tracks and collisions of every single gas molecule inside the layer. Molecular collisions are calculated by employing collision parameters acquired from simplified solutions of the Boltzmann equation. If a model of the porous geometry to investigate is available, the DSMC method is therefore able to accurately describe gas diffusion processes in inhomogeneous layers, without the knowledge of pore diameter, tortuosity, or porosity information. The OpenFOAM implementation of the DSMC code has been extended by the Variable Soft Sphere (VSS) model for binary molecular collisions [1], which results in a more accurate diffusion rate, compared the Variable Hard Sphere (VHS) model. Further, diffusion of CO into a N_2 filled layer has been simulated by the DGM and the DSMC-VSS code. Whereas DSMC and DGM show a good agreement for the diffusion inside isotropic layers, DSMC shows higher accuracy for diffusion inside real gas sensor layers as obtained by an aerosol synthesis method.

To obtain information about the chemical reaction fronts inside the layer, the DSMC method has been extended to describe basic heterogeneous reaction mechanisms, i.e., adsorption, co-adsorption, desorption and reaction of gas species on the surface of the solid [2]. The adsorption is based on the well-known sticking coefficient and implemented by the Kisluik model for precursor mediated adsorption, which describes the chance of a molecule to stick on the surface after hitting the solid as a function of temperature and coverage of the surface. Desorption and surface reaction are modeled through a mean-field rate approximation, i.e., the Polanyi-Wigner equation for desorption and a second-order Arrhenius equation for reaction. With this model we study the catalytic oxidation of carbon monoxide (CO) inside gas sensor layers of 1000 nm thickness in the transition regime ($Kn \sim 1$) using kinetic parameters taken from the literature for single-crystal Pd(111) surfaces at UHV conditions (Fig 1a).

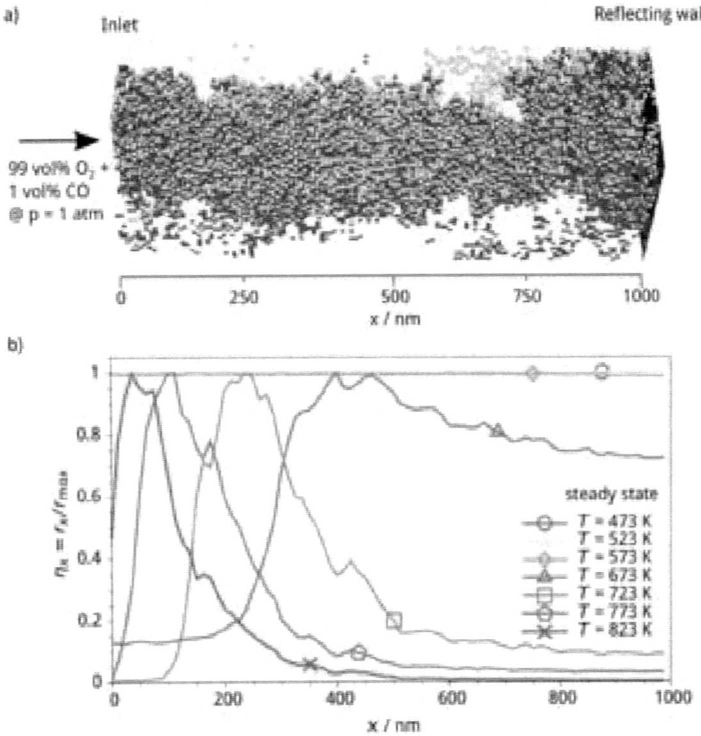

Fig. 1: a) Investigated porous layer and simulation parameters. b) Normalized reaction rate η_x as a function of the penetration depth in the layer x at 7 different temperatures T.

Investigation of the reaction at different temperatures reveals a clear transition from a kinetic limit at low temperatures ($T < 673$ K) to a diffusion limit at high temperatures ($T > 673$ K). In the kinetic limit, the diffusion occurs at much faster rate than the reaction; the latter is therefore the rate-determining step of the overall process. At the diffusion limit, the reaction consumes educts at a much higher rate than their replenishment due to diffusion from the inlet. At this limit, diffusion of educts is thus the rate-determining step and the process is mass-transport limited.

At high temperatures and at steady state (Fig 1b), the layer is separated into three distinct regions. The surface of the layer is poisoned by CO, which is a result of the underlying co-adsorption mechanism. Here, the surface is covered with CO, which is hindering oxygen adsorption (competitive adsorption mechanism). Hence, the reaction rate is low, as only one out of two species (CO) required for the reaction is adsorbed on the surface.

With increasing depth into the layer the CO coverage is decreasing and the oxygen coverage is increasing, which results in an effective reaction area of the layer. The peak reaction rate is reached as both coverages intersect. Going even deeper into the layer, the CO coverage reaches zero, as the effective reaction area consumes all CO. This reflects the mass-transport limitation of the overall reaction. Hence, the reaction rate is again low as only one out of two species needed for the reaction (oxygen) is adsorbed on the surface.

The employed parameters together with the presented reaction system serve to demonstrate the capabilities of the newly developed simulation algorithm. We expect that similar investigations will not only help gaining deeper understanding of the reaction processes inside porous structures but also for the structure optimization of gas sensors and catalysts.

[1] J.A.H. Dreyer, N. Riefler, G.R. Pesch, M. Karamehmedovic, U. Fritsching, W.Y. Teoh, L. Mädler: Simulation of Gas Diffusion in Highly Porous Nanostructures by Direct Simulation Monte Carlo. Chem. Eng. Sci. (2014), 69-76

[2] G.R. Pesch, N. Riefler, U. Fritsching, L. Colombi Ciacchi, L. Mädler: Gas-Solid Catalytic Reactions with an Extended DSMC Model. AIChE J. (2015), in press, DOI: 10.1002/aic.14856

427218 A Numerical Optimization Study on the Catalytic Dry Reforming of Methane in a Spatially Resolved Fixed-Bed Reactor

Abstract:

A Numerical Optimization Study on the Catalytic Dry Reforming of Methane in a Spatially Resolved Fixed-Bed Reactor

Thomas Eppinger[1], Nico Jurtz[1], Gregor D. Wehinger[2], Matthias Kraume[2], Ravindra Aglave[3]

[1] CD-adapco, Nuremberg, Germany

[2] CD-adapco, Houston,TX

[3] Chair of Chemical and Process Engineering, Technische Universität Berlin, Germany Introduction

Fixed bed or packed bed reactors are widely used in the chemical and process industry amongst others for highly exothermic or endothermic catalytic surface reactions. A small tube to particle diameter ratio (D/d-ratio) facilitates a safe thermal performance of these reactors. Many correlations reported in literature for pressure drop and heat transfer are limited in their predictive capability for such packed bed reactors due to a dominant influence of the confining wall. It affects the porosity distribution and as a result the velocity field as well as the species and temperature distribution within the bed. Accurate measurements for development of new correlations can be cumbersome, expensive and many times inaccurate due to the complexity of the geometry. The increasing power of computational hardware in the last decade has enabled virtual experiments: Computational Fluid Dynamics (CFD) can now be used for three dimensional spatially resolved investigations of those reactor types. Aim of this contribution

The current contribution is such an investigation of the catalytic dry reforming of methane (DRM) in a packed bed reactor based on the detailed reaction mechanism proposed by [McGuire et al., 2011]. It will be shown how a fully automated workflow for a three dimensional and spatially resolved packed bed simulation with rigorous optimization methods can lead to a significant increase in terms of maximum yield and space-time yield. Methods

Discrete Element Method (DEM) is used for the generation of a randomly packed bed. A finite volume (FV) code is used for the calculation of the fluid flow, temperature and species field within the packed bed. To avoid the well-known meshing problems near the particle-particle and particle-wall contacts, a local wall flattening approach is used.

These methods have been validated in terms of packing structure, pressure drop, velocity and temperature profiles ([Eppinger et. al., 2011] and [Zobel et. al., 2012]) and were recently used to investigate the effect of different particle shapes on the performance of a packed bed reactor ([Wehinger et al., 2015]).

In this study a detailed surface reaction mechanism [McGuire et al., 2011] is used and solved by DARS-CFD, a code that is developed to solve stiff differential equation systems efficiently. For the optimization of operating parameters the SHERPA algorithm of HEEDS by Red Cedar Technology is used. SHERPA is a hybrid, adaptive optimization algorithm that blends global and local optimization methods to provide the most efficient route to an optimized design. The whole analysis is performed using STAR-CCM+ by CD-adapco which allows an out of the box coupling of the FV code with DARS (using DARS-CFD) and HEEDS (using the Optimate/Optimate+ plug in). Results

The sensitivity of the yield and space-time to parameters like residence time, inlet and wall temperature is calculated. The working parameters are optimized to maximize the yield while taking the constraints such as maximum local temperature into account. The results of the spatially resolved simulation will finally be compared with results of a pseudo-homogenous (unresolved in 3D) simulation and published experimental data. References

N. E. McGuire, N. P. Sullivan, O. Deutschmann, H. Zhu, R. J. Kee *Dry reforming of methane in a stagnation-flow reactor using Rh supported on strontium-substituted hexaaluminate*, Applied Catalysis A: General. 2011, 257-265

T. Eppinger, K. Seidler, M. Kraume. *DEM-CFD simulations of fixed bed reactors with small tube to particle diameter ratios*. Chemical Engineering Journal 2011, 166(1): 324-331.

G. D. Wehinger, T. Eppinger, M. Kraume. *Evaluating Catalytic Fixed-Bed Reactors for Dry Reforming of Methane with Detailed CFD*, Chem. Ing. Tech., in press

N. Zobel, T. Eppinger, F. Behrendt, M. Kraume. *Influence of the wall structure on the void fraction distribution in packed beds*. Chemical Engineering Science 2012, 212–219.

444576 Molecular-Dynamics-Assisted Molecular-Thermodynamic Framework to Predict the Interfacial Tensions of Non-Ionic Surfactants

Abstract:

The reduction in interfacial tension by surfactants underlies several natural phenomena in multi-phase systems, including emulsions and foams. This effect is also responsible for many industrial processes such as spray painting, emulsion polymerization, distillation in packed bed columns, and froth flotation. The mechanisms underlying these phenomena have been described by several mathematical models in the colloid science literature. Since the interfacial tension is an essential input for the implementation of all these models, their accuracy depends on the accuracy of the estimate (s) of the interfacial tension (s) used. Here, we propose a method to evaluate the surface tensions of non-ionic surfactants using a combination of molecular dynamics (MD) simulations and molecular-thermodynamic (MT) theory. We have developed an automated, computational workflow to carry out a series of simulations of surfactant molecules at fluid-fluid interfaces. One set of simulations studies the interactions between surfactant molecules, and attempts to characterize the intermolecular interactions in a surfactant monolayer in terms of two molecular parameters – a hard-disk radius (representative of strong intermolecular repulsions) and a second-virial coefficient (representative of weak, intermolecular, van der Waals attractions). These parameters are obtained by first calculating the potential of mean force between the two surfactant molecules at the interface, and then analyzing this potential using principles of statistical mechanics. A second set of simulations studies the change in the free energy of a surfactant molecule as it is brought from bulk solution to an interface. This free energy of adsorption is a measure of the affinity of the surfactant of interest for the interface of a given system, and provides a metric capable of ranking families of surfactants. Together with the intermolecular interaction parameters, the free energy of adsorption enables prediction of the interfacial tension as a function of the surfactant bulk concentration through the use of a suitable equation of state for the interface.

In this talk, we will present the details of our proposed computational methodology, including: (1) the results obtained by its application to a series of nonionic alkyl poly (ethylene oxide) (CiEj) surfactants at fluid-fluid interfaces, and (2) our perspective on applying this methodology to high-pressure/high-temperature systems.

425211 Development of a Comprehensive Molecular Thermodynamic Model for High Salinity Produced Water in Oil and Gas Productions

Abstract:

Development of a Comprehensive Molecular Thermodynamic Model for High Salinity Produced Water in Oil and Gas Productions

Chau-Chyun Chen

Department of Chemical Engineering, Texas Tech University, Lubbock, TX 79409-3121, USA

Abstract

Managing produced water in the rapidly developing shale gas industry is a major challenge for protecting the environment [Shaffer et al., *Environ. Sci. Technol.* 2013, 47, 9569-9583]. Desalination for reuse of produced water depends on novel material and process innovations. We present progress on the development of a comprehensive thermodynamic model aimed to support heat and mass balance calculations and process simulation of desalination processes with produced water. Based on the electrolyte Non-Random Two Liquid theory (eNRTL) for electrolyte solutions [Song and Chen, *Ind. Eng. Chem. Res*. 2009, 48, 7788-7797], the model intends to cover major ions of concern for produced water management: Na^+, K^+, Mg^{2+}, Ca^{2+}, Ba^{2+}, Sr^{2+}, Cl^-, HCO_3^- and SO_4^{2-} ions. In the thermodynamic framework of eNRTL, the concentration dependency of the solution nonideality is accounted for with two binary interaction parameters for each of the water-electrolyte interaction pairs and the electrolyte–electrolyte interaction pairs. The temperature dependency of the binary interaction parameters is accounted for with a Gibbs-Helmholtz type expression with three temperature coefficients representing excess Gibbs energy, excess enthalpy, and excess heat capacity contributions. With the binary parameters regressed from thermodynamic data of aqueous single electrolyte binary systems and aqueous two electrolyte ternary systems, the model has been shown to provide accurate calculations and reliable predictions for various thermodynamic properties of quaternary and quinary systems examined thus far for the aqueous Na^+-K^+-Mg^{2+}-Ca^{2+}-Cl^--SO_4^{2-} hexary system with temperatures from 273.15 K to 473.15 K and concentrations up to salt saturation. The model is being extended to cover Ba^{2+}, Sr^{2+} and HCO_3^- ions and it should become an indispensable scientific tool in the development of novel desalination processes for high salinity produced water in oil and gas productions.

426263 Gas Hydrate Phase Behavior in Presence of Organics and Electrolytes

Abstract:

In this contribution, the effects of electrolytes on the phase behaviour of carbon dioxide hydrate forming systems are presented. The phase behaviour of the simple carbon dioxide hydrate system, the mixed carbon dioxide - tetrahydrofuran (THF) hydrate system, and, finally, the effect on the phase behaviour of different overall concentrations of sodium chloride will be elucidated. In addition, the competing effects between THF and an electrolyte of the metal halide series and their impacts on the phase behaviour of the hydrate forming systems will be discussed. The strength of hydrate inhibition by metal halides is compared and, a quantitative analyses between the effect of anions and cations is made in order to gain some understanding on the mechanism of electrolyte inhibition on hydrate formation.

It is well established that electrolytes inhibit clathrate hydrate formation. The inhibition of hydrate formation by electrolytes is due to the lowering of the activity of water in the coexisting liquid phase, causing hydrates to form at lower temperatures and higher pressures compared to their formation in pure water [1]. Many phase equilibrium data especially on H-L_W-V equilibrium line have been reported for carbon dioxide in the presence of electrolytes. A compilation of these data can be found in Sloan [2]. In contrast, cyclic organics such as tetrahydrofuran reduce the pressure requirement for hydrate formation at a specified temperature when presents in small quantity in aqueous solutions up to about 5 – 7 mol%. The usage of such organics at low concentration may provide an alternative solution to overcome the inhibiting effects of electrolytes. Unfortunately, no available reported data for hydrate equilibria of mixed carbon dioxide and tetrahydrofuran hydrates in aqueous solution have been found in literature. Realizing the potential benefits of such data for the development of hydrate-based processes, an attempt has been made to measure the hydrate equilibrium data for such a system. For consistency, tetrahydrofuran is chosen as a representative of hydrate promoter in the present work. In this study, phase behaviours of mixed carbon dioxide and tetrahydrofuran hydrates in sodium chloride (NaCl) solutions are presented. Sodium chloride is chosen to represent electrolytes that are present in naturally occurring waters. The measurement of hydrate equilibrium conditions for mixed carbon dioxide and tetrahydrofuran hydrates are extended to another six electrolytes from the metal halides series. Besides NaCl, the electrolytes used in this work are calcium chloride ($CaCl_2$), magnesium chloride ($MgCl_2$), potassium bromide (KBr), sodium fluoride (NaF) and sodium bromide (NaBr) and potassium chloride (KCl). Based on measured data, a comparison is made to quantitatively evaluate the effect of different salts inhibition on hydrate formation.

In the quaternary system studied, the presence of electrolyte in the system enhanced the liquid-liquid phase split creating two distinct liquid phases of liquid water and liquid organics (tetrahydrofruan + carbon dioxide). This behaviour is attributed to the salting-out effect that taking place in the system due to the presence of ions. In general, it is found that the hydrate promoting effect of tetrahydrofuran suppresses the inhibiting effect of an electrolyte when both are present in a hydrate forming system. The strength of hydrate inhibition effect among the electrolytes was compared. The results shows the hydrate inhibiting effect of the metal halides is increasing in the order NaF < KBr < NaCl < NaBr < $CaCl_2$ < $MgCl_2$. Among the cations studied, the strength of hydrate inhibition increases in the following order: $K^- < Na^- < Ca^{2+} < Mg^{2+}$. Meanwhile, the strength of hydrate inhibition among the halogen anion studied decreases in the following order: $Br^- > Cl^- > F^-$. Based on the results, it is suggested that the probability of formation and the strength of ionic-hydrogen bond between an ion and water molecule and the effects of this bond on the ambient water network are the major factors that contribute to hydrate inhibition by electrolytes.

References

[1] Z. Duan, R. Sun, American Mineralogist, 91 (2006) 346-1354.

[2] E.D. Sloan Jr., C.A. Koh, "Clathrate Hydrates of Natural Gas", 2nd ed., CRC Press, Florida, 2008.

417023 Phase-Portrait of Complex Kinetic Systems

Abstract:

Phase-Portrait of Complex Kinetic Systems

Session 01A00

Salt Lake City, Utah

Nov. 8-13, 2015

James Wei, Princeton University

In a complex kinetic system with many components, the phase-portrait is the global qualitative description of the totality of the reaction paths of the system. It can be used to analyze and to optimize selectivity and product values. The angle of the slow eigenvector and the direction for the convergence of reaction paths are dominant for the best operational and research strategies.

417722 Phase Equilibria of Ternary Mixtures of Two Nematic Liquid Crystals and CO2 for CO2 Capture

Abstract:

Liquid crystals are rod-like molecules, consisting of a rigid core and flexible tail, with a structured phase between the usual liquid and solid phase which is still fluid. This structured phase is called nematic if the molecules are orientationally ordered. Currently, liquid crystals are examined as novel solvents for capturing CO_2.[1,2] As CO_2 is less soluble in the structured phase than in the isotropic phase, gaseous CO_2 can be separated from the structured phase by cooling down a saturated isotropic mixture a few degrees at constant pressure. The solvent can be regenerated by heating the CO_2 lean mixture a few degrees again.

To obtain the optimal liquid crystal for the CO_2 capture process, binary mixtures of a liquid crystal and CO_2 have been examined with different concentrations of CO_2. Liquid crystals examined had either two phenyl rings, two cyclohexyl rings or one phenyl and one cyclohexyl ring. We can conclude, based on experimental results, that molecules with both a cyclohexyl and phenyl ring have highest solubility. The solubility difference between the isotropic and nematic phase is mainly determined by the phase transition enthalpy of the isotropic-nematic phase transition.

However, no molecule has been found yet with optimal properties. Therefore, mixtures of two liquid crystals and CO_2 are investigated.[3] Experiments showed that in binary mixtures of two liquid crystals, the nematic phases form a nematic homogeneous solution and the solid phases form an eutectic system. This leads to a solvent with improved properties for CO_2 capture, as the temperature range in which the nematic phase is stable is increased, leading to an increased solubility difference of CO_2 between the nematic and isotropic phase. The ternary mixtures 4,4'-pentyloxycyanobiphenyl + 4,4'-heptyloxycyanobiphenyl + CO_2 and 4,4'-propylcyclohexylbenzonitrile + 4,4'-heptylcyclohexylbenzonitrile + CO_2, were investigated. The ternary mixture of 4,4'-propylcyclohexylbenzonitrile + 4,4'-heptylcyclohexylbenzonitrile + CO_2 showed an increased solubility of CO_2 compared to the binary subsystems.

Furthermore, mixtures can be used for fine-tuning the phase transition temperature.

References

[1] Gross, J. Jansens, P.J., A method for the removal of a gas from a process gas stream by means of liquid crystals, 2008, Patent WO2008147181-A1; NL2000654-C2.

[2] de Groen, M. Vlugt, T.J.H. de Loos, T.W., Phase behavior of liquid crystals with CO_2, J. Phys. Chem. B, 2012, 116 (30), 9101-9106.

3 de Groen, M. Vlugt, T.J.H. de Loos, T.W., Binary and Ternary Mixtures of Liquid Crystals with CO_2, AIChE journal, accepted

15AIChE Annual Meeting, Salt Lake City, UT

In Honor of Stanley Sandler – (74e)

Effect of additives on hydrate formation

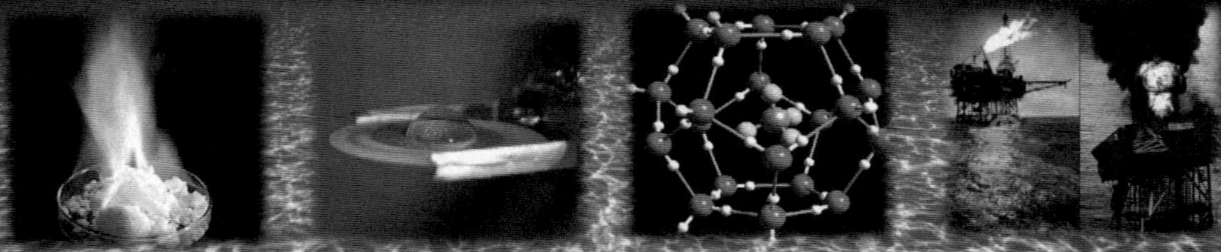

Jeong-Hoon Sa,[1] Bo Ram Lee,[2] Da-Hye Park,[3] Gye-Hoon Kwak,[1] Kunwoo Han,[4] Kun-Hong Lee[1]*

[1]Department of Chemical Engineering, Pohang University of Science & Technology (POSTECH), Korea
[2]Chemical and Biological Engineering Department, Colorado School of Mines, USA
[3]Energy Laboratory, Samsung Advanced Institute of Technology, Korea
[4]CO_2 Project Team, Research Institute of Industrial Science & Technology, Korea

9th Nov. 2015

POSTECH

A little background

- Kun-Hong Lee
  - In honor of Stanley Sandler's 70th Birthday : Electrochemical Porosimetry
  - In honor of Stanley Sandler's 75th Birthday : Gas Hydrates

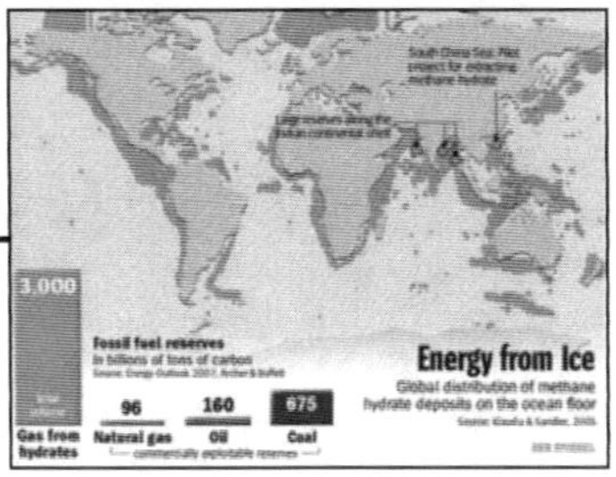

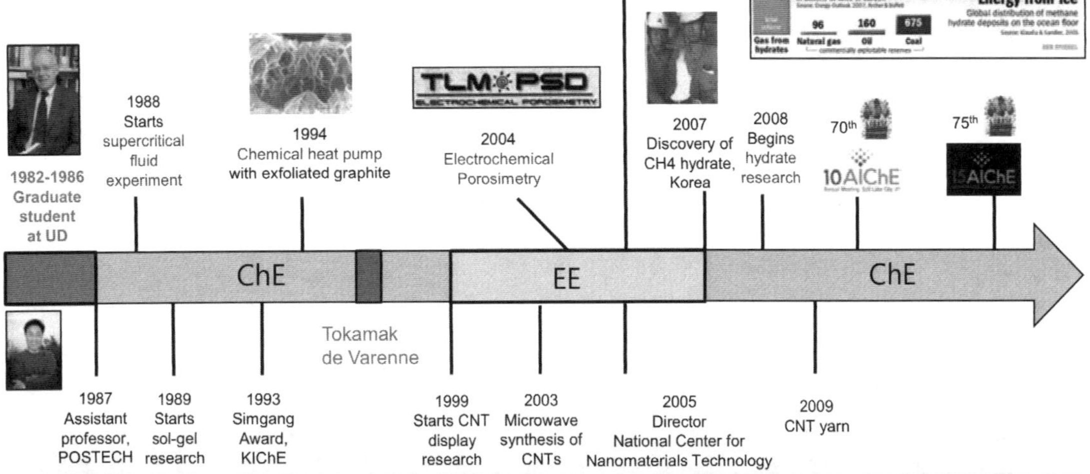

Gas hydrates

- Crystalline ice-like solid materials, in which gas molecules are enclathrated in a framework linked by hydrogen-bonded water molecules

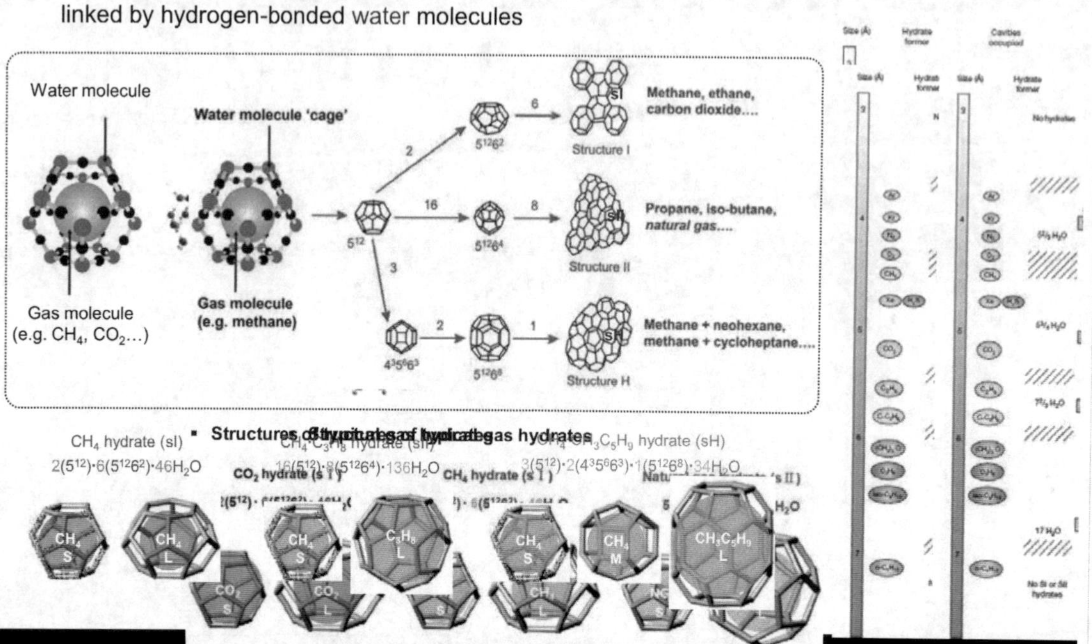

Nature **2003**, *426*, 353. *Annu. Rev. Mar. Sci.* **2009**, *1*, 303. http://www.pet.hw.ac.uk/research/hydrate/hydrates_what.cfm?hy=what. 3

Hydrate additives

- Thermodynamic effects: Hydrate phase equilibria
- Kinetic effects: Hydrate nucleation and/or growth rate

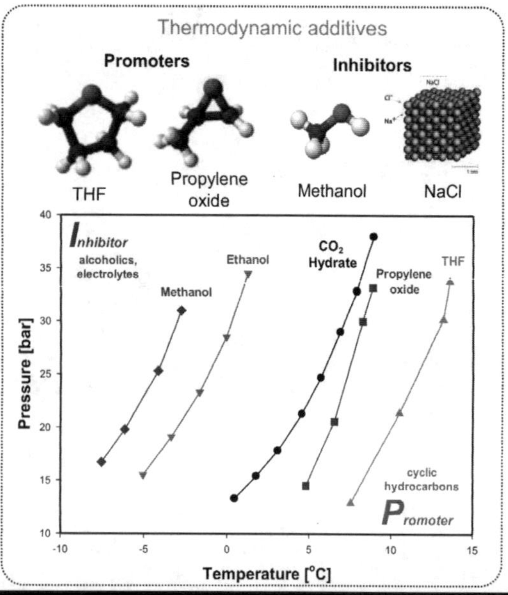

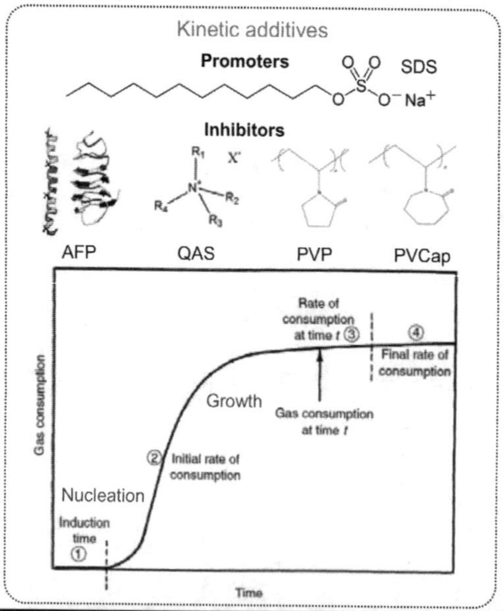

E. D. Sloan, C. A. Koh, *Clathrate Hydrates of Natural Gases*, 3rd ed., CRC Press, FL, 2008. 4

Contents

❑ Promotors : Quick screening method

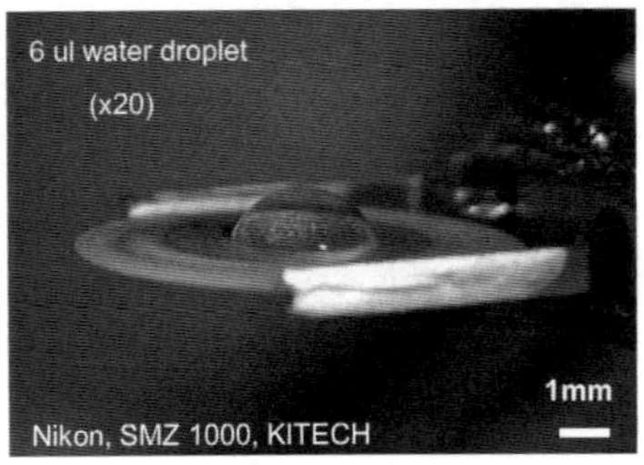

❑ Inhibitors : Amino acids

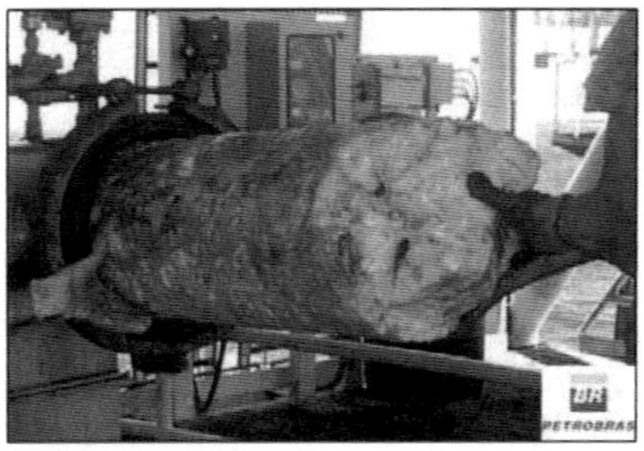

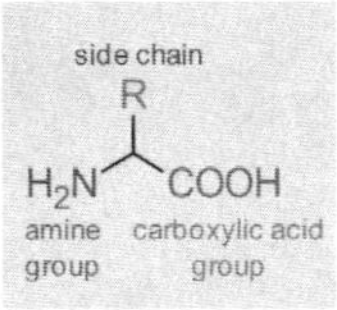

Screening of additives

- The selection of more efficient additives
  - Problems: The time required for additive evaluation — Too slow to measure
    A variety of additives — Too many candidates

Too long time !
(24 hours / 1 point)

Heating rate : 0.1℃ / 1hour
5~6 point required
+ Error bar

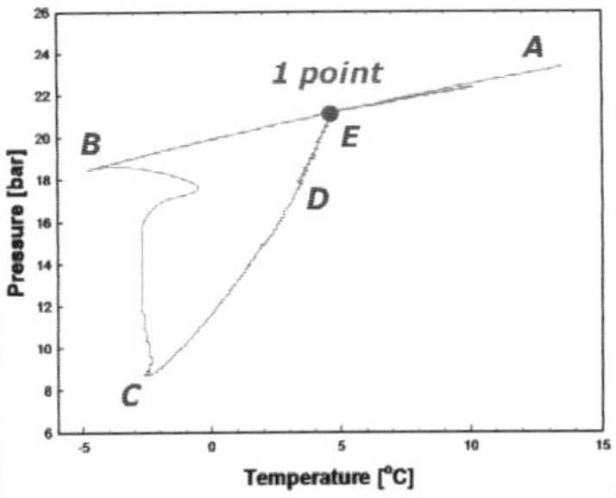

How ?

Any quick screening tool !?

Micro DSC, SETARAM
But, expensive…

A lot of Target materials !
(Promoters & Inhibitors)

Organic
Inorganic **materials**
Gaseous

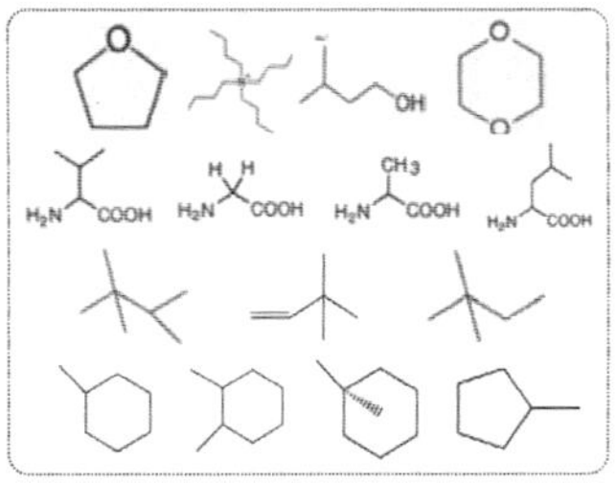

※ Millions of target materials

http://img.directindustry.com/images_di/photo-g/differential-scanning-microcalorimeters-dsc-271556.jpg12

QCM method

- Quartz Crystal Microbalance (QCM)
  - Piezoelectricity, extreme sensitivity
  - Measuring the gas hydrate stability zone based on the change in the frequency

QCM — QCM pressure cell

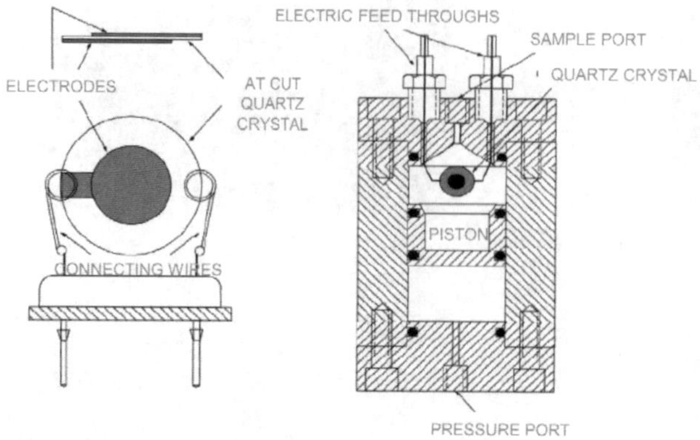

Introduced by Prof. Tohidi in 2002

"QCM method requires that hydrates have to adhere to the surface of the quartz crystal, and in some cases this may not occur, thereby making these measurement unfeasible."

(*Clathrate Hydrates of Natural Gases*. 333p)

J. Chem. Eng. Data **2003**, *48*, 612-626. Tohidi, US Patent 6,298,724 B1 (2001) *Chem. Eng. Sci.* **2005**, *60*, 4203-4212. 7

Problem

- Experimental procedure

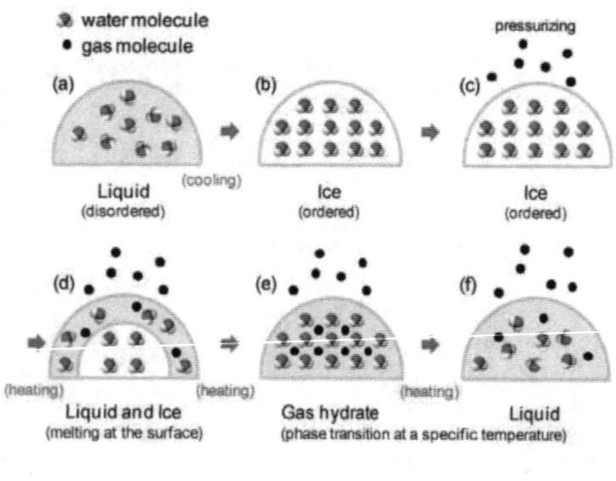

- Hydrate film forms on the surface of a water droplet.
- Gas diffusion through the solid hydrate film is slow.

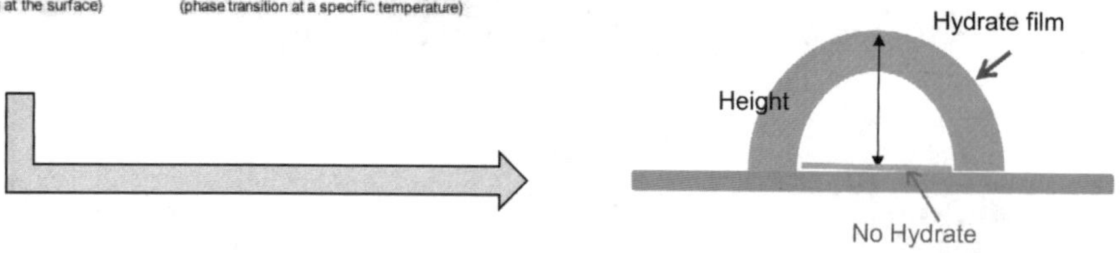

Solution

- Reduce the drop size

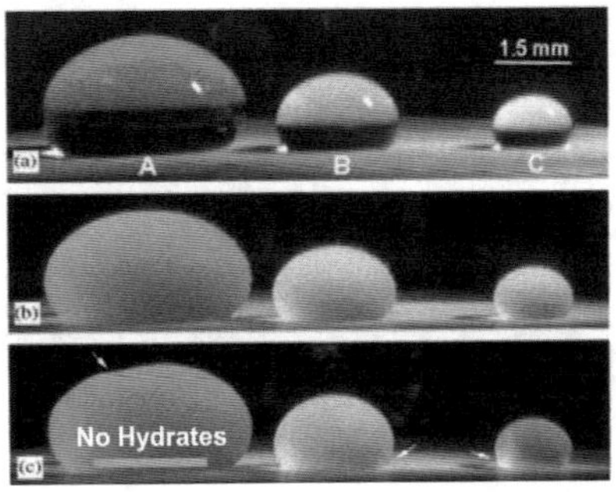

Size control (2~5μℓ droplet)

- Quartz crystal surface treatment

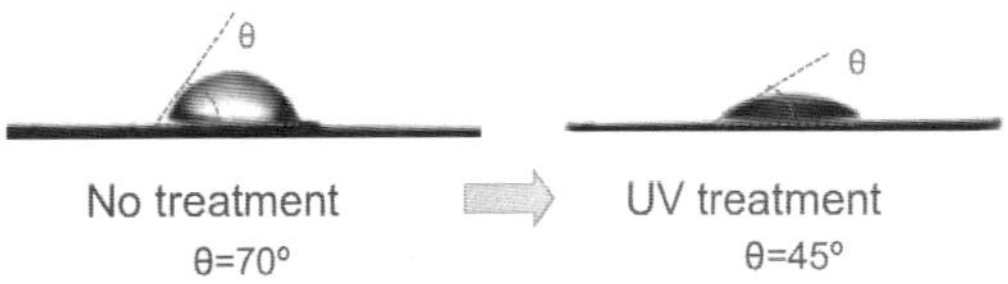

All surface, no flesh !

Contact between hydrate film and quartz crystal is almost guaranteed.

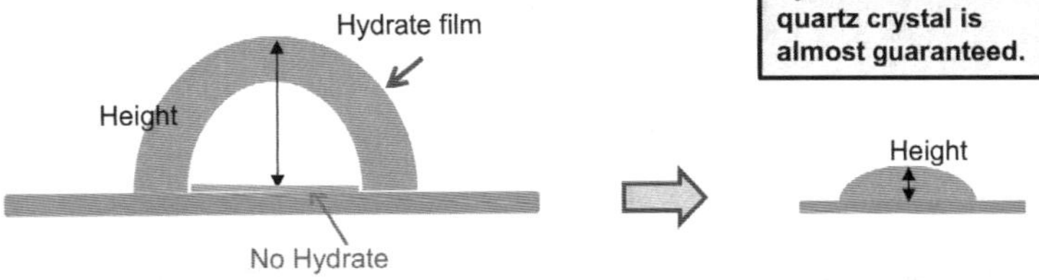

Result

- Fast measurements were possible.

	Conventional cell	QCM cell
Water volume	100 cc	0.004 cc
Measuring time	24 h	2 h

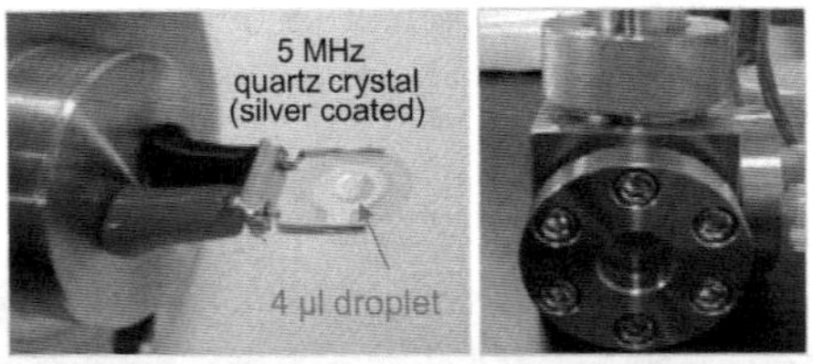

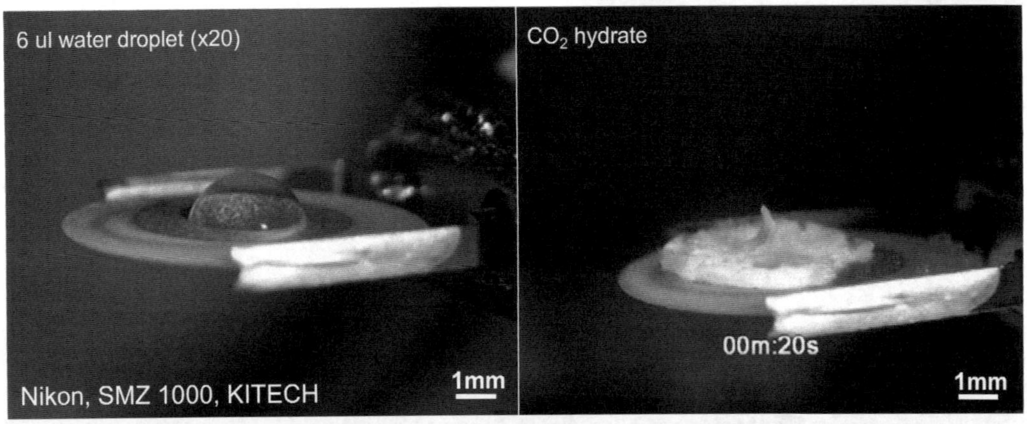

10

Quick screening method

❖ New QCM method

- Frequency & resistance change with phase transition

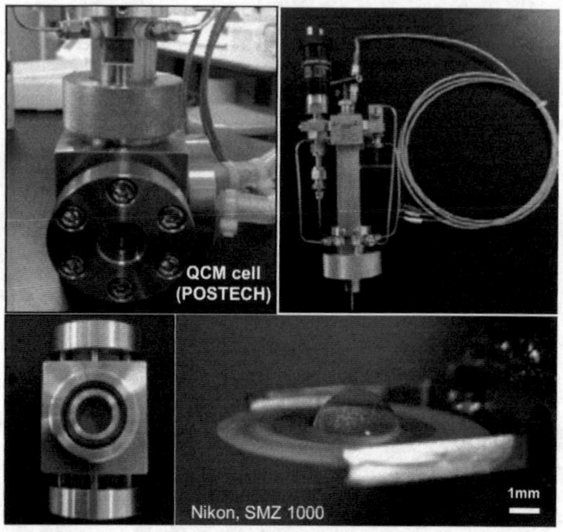

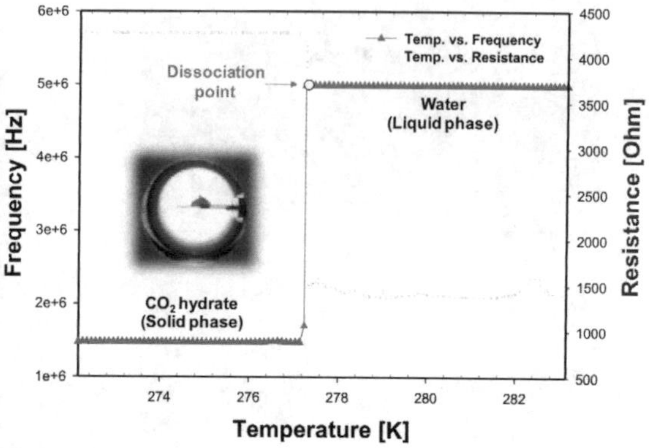

- Max pressure: 70 bar
- Reactor volume: **11 cc**
- Quartz crystal : **5 MHz** silver coated
- QCM-200 : Data acquisition

Lee et al. *Energy Fuels* **2012**, *26*, 767-772. Korea Patent KR 10-1255753 (2013.4.11.)

Reliability

❖ Excellent agreement with the data from literature

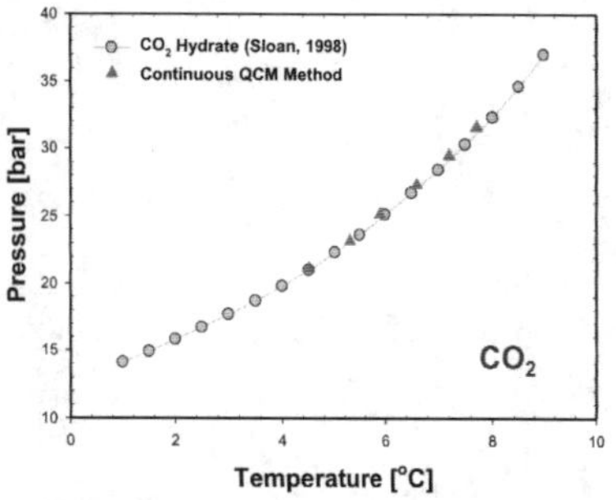

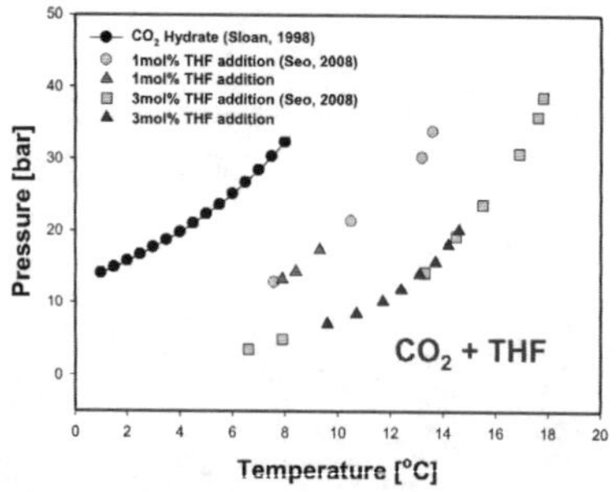

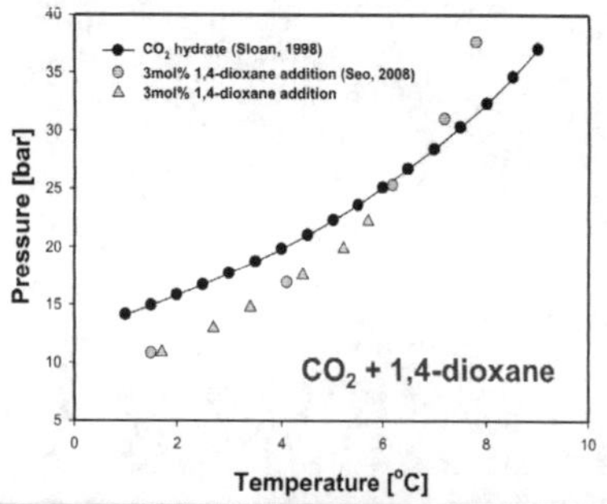

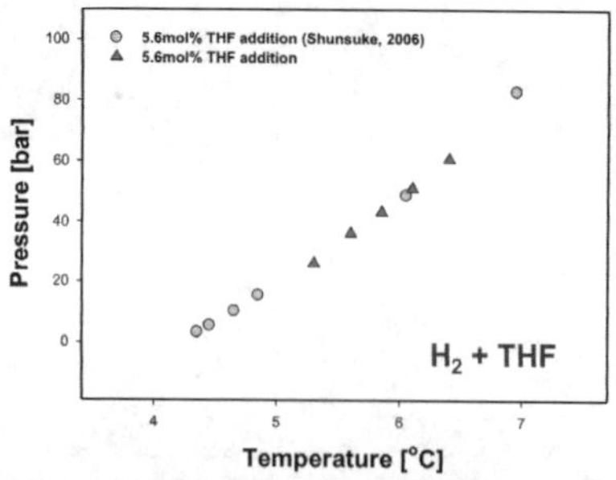

Lee et al. *Energy Fuels* **2012**, *26*, 767-772. Patent registration KR 10-1255753 (2013.4.11.)

Multi-QCM system

- Continuous measurement using the "memory effect"
  - Hydrates form more easily with previous hydrate history.
  - Quick measurement: 24 h/point (Conventional method)
    1 h/point ("Continuous" method)

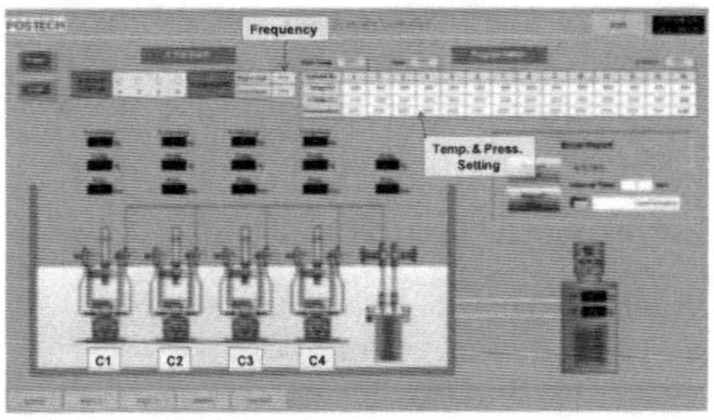

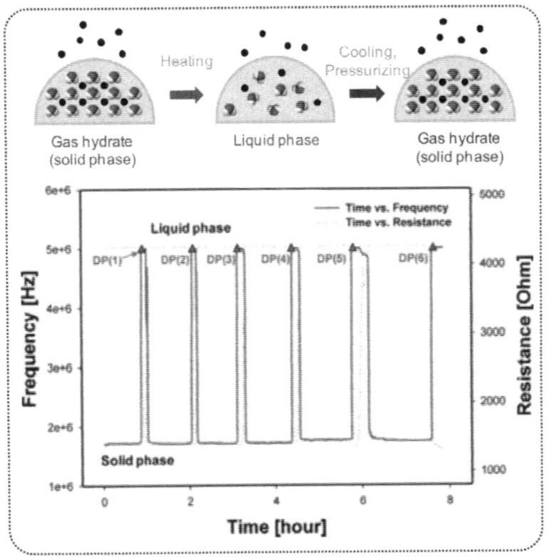

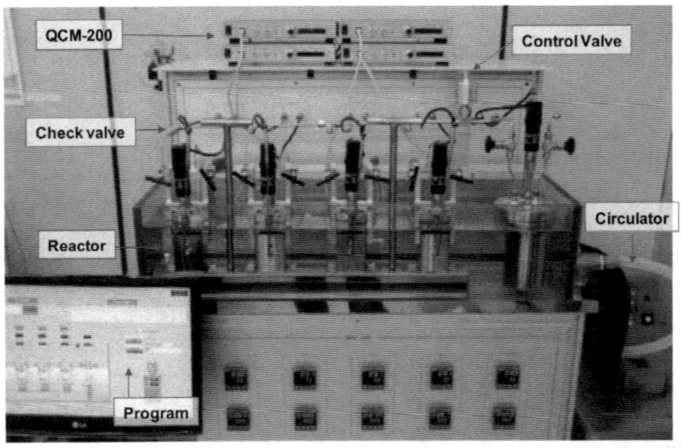

Lee et al. *Energy Fuels* **2012**, *26*, 767-772. Patent registration KR 10-1255753 (2013.4.11.) 13

Inhibitors for flow assurance

- Inhibitors used for flow assurance are potentially harmful to the environment.
- Fish in the antarctic water have antifreeze protein in the blood.

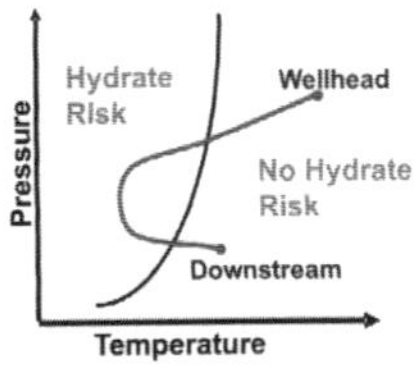

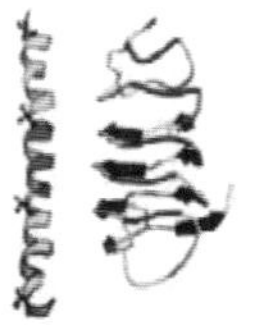

AFP

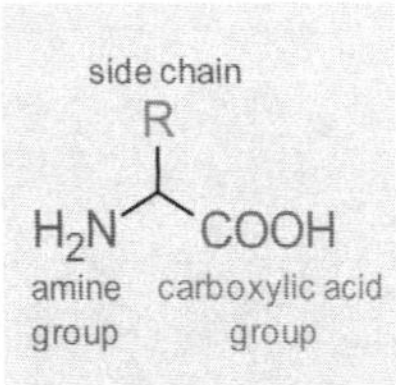

Amino acid

- How about amino acid ?
  - Naturally occurring molecules with inherent environmentally friendly nature
  - Containing carboxylic acid group (-COOH) & amine group ($-NH_2$) *"hydrogen bond"*
  - Spontaneous zwitterion formation in water *"electrostatic interaction"*
  - Containg hydrophobic side chain with different sizes
  - A good academic model system

J. Mol. Biol. **1982**, *157*, 105. P. Y. Bruice, *Organic Chemistry*, 5th ed., 2007. 14

Amino acids: unique inhibitor

❖ Amino acids: promising dual function inhibitors (THIs & KHIs)

✓ Hydrophobic amino acids: Hydrophobicity depends on the alkyl chain length.

Molecular structure	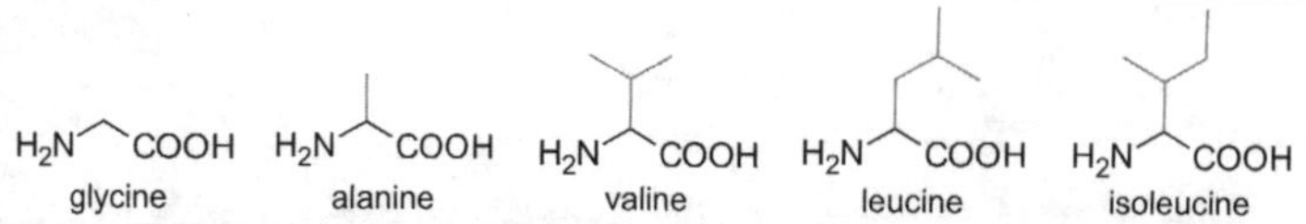 glycine	alanine	valine	leucine	isoleucine
Side chain	-H	$-CH_3$	$-CH(CH_3)_2$	$-CH_2CH(CH_3)_2$	$-CH(CH_3)C_2H_5$

✓ Amino acids with various side chains: A side chain determines the physicochemical property.

Molecular structure	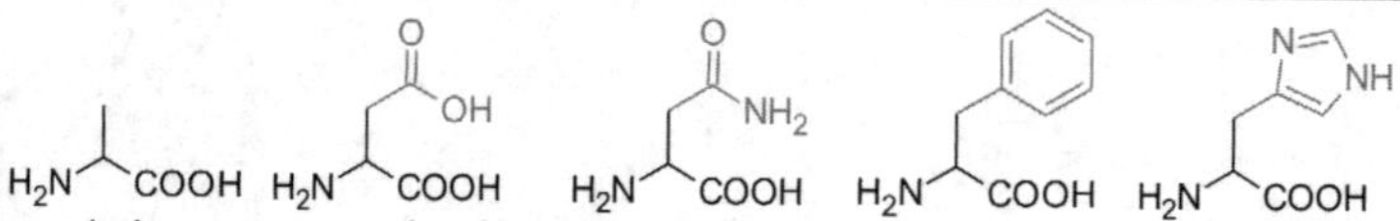 alanine	aspartic acid	asparagine	phenylalanine	histidine
Side chain	$-CH_3$	$-CH_2COOH$	$-CH_2CONH_2$	$-CH_2C_6H_5$	$-CH_2C_3H_3N_2$

J. Mol. Biol. **1982**, *157*, 105. P. Y. Bruice, *Organic Chemistry*, 5th ed., 2007. 15

Thermodynamic inhibitors

❖ CO_2 hydrate phase equilibria

- Hydrophobic amino acids, including glycine, L-alanine, and L-valine have abilities to shift the CO_2 hydrate formation conditions to lower temperature and higher pressure region.
- Amino acids with shorter alkyl chains (lower hydrophobicity) were more useful as much soluble in water.

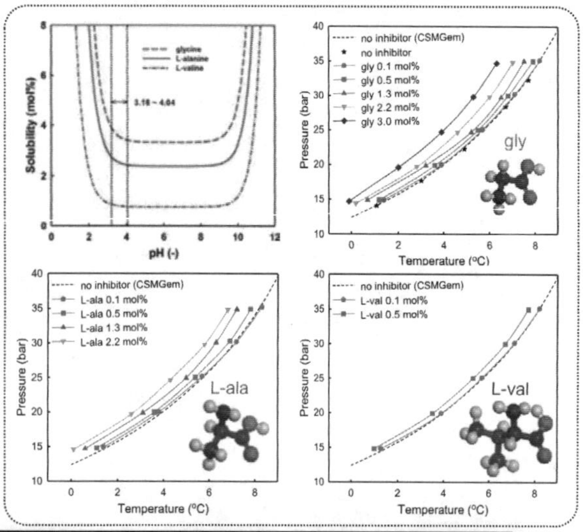

On the maximum concentration basis,

glycine > L-alanine > L-valine

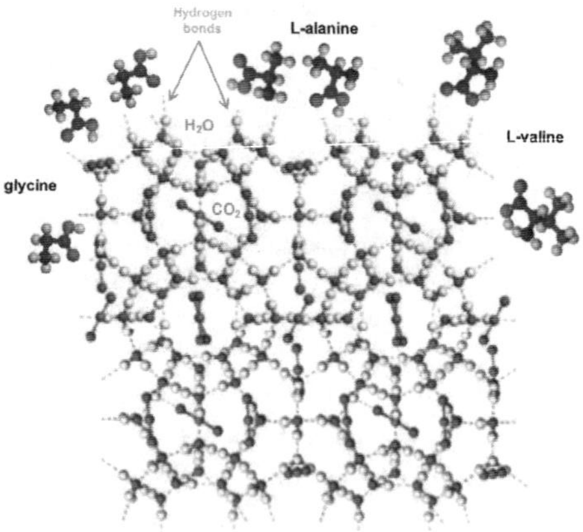

Sa et al. *Environ. Sci. Technol.* **2011**, *45*, 5885-5891. Paten

Hydrophilic amino acids

❖ CH_4 & natural gas hydrate phase equilibria

- Amino acids, including glycine, L-alanine, L-serine, and L-proline have abilities to shift both CH_4 and natural gas hydrate formation conditions to lower temperature and higher pressure region.
- Amino acids with lower hydrophobicity were more useful as much soluble in water.
- The 9.0 mol% injection of L-proline even shifted the formation conditions of NG hydrates to those of pure CH_4 hydrates.

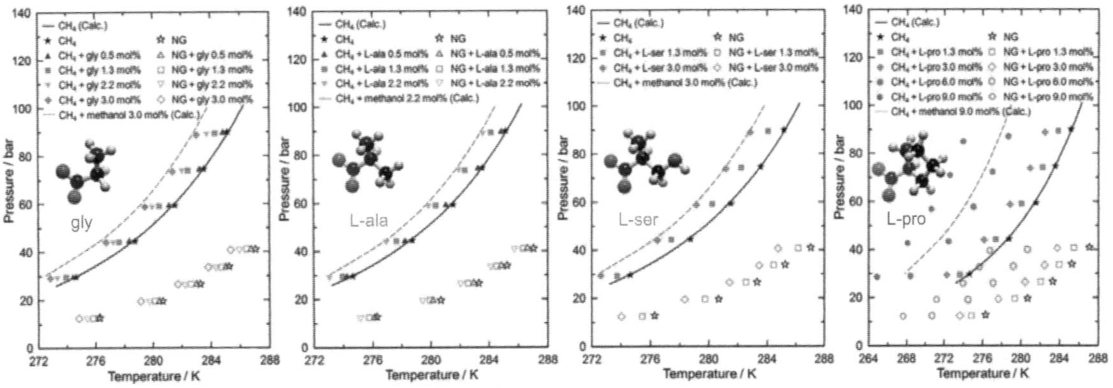

Sa et al. *Submitted.*

Kinetic inhibitors

❖ Subcooling temperature

- Two hydrophobic amino acids (gly and L-ala) have abilities to delay CO_2 hydrate nucleation.
- Amino acids with shorter alkyl chains (lower hydrophobicity) were more efficient KHIs, especially in memory water (with hydrate history).

Stage 1. Onset of hydrate formation

Concentration dependence

Alkyl chain length (hydrophobicity) dependence

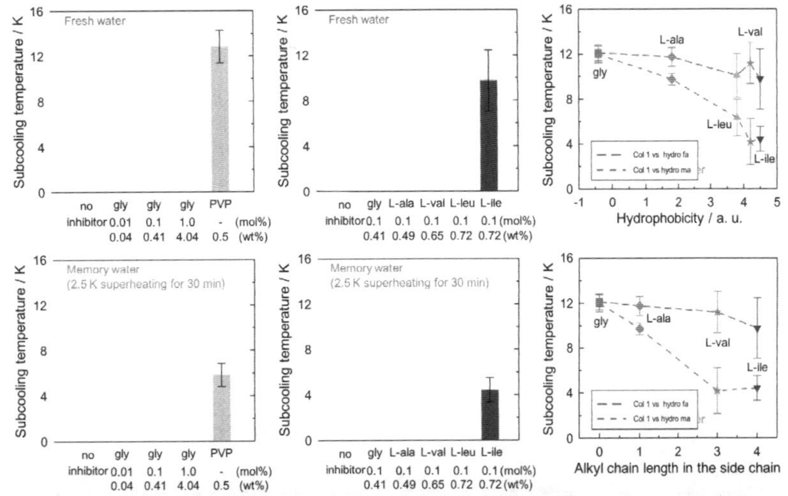

Sa et al. *Sci. Rep.* **2013**, *3*, 2428.

Growth kinetics

❖ Gas uptake rate & Synchrotron PXRD patterns

- Two hydrophobic amino acids (gly and L-ala) have abilities to retard CO_2 hydrate growth.
- As the gly concentration increased, intensities of the diffraction peaks of I_h phase increased.

Stage 2. During hydrate formation *Stage 3. After hydrate formation*

Concentration dependence

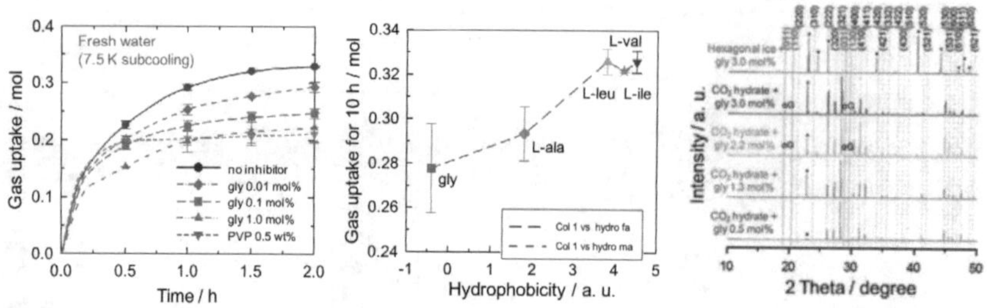

Alkyl chain length (hydrophobicity) dependence

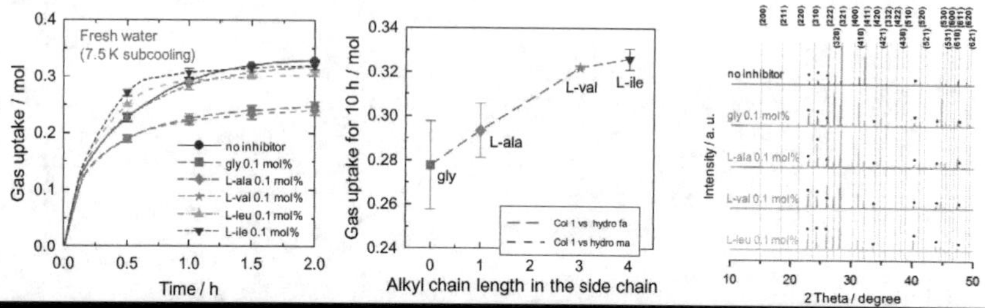

Sa et al. *Sci. Rep.* **2013**, *3*, 2428. Sa et al. *Phys. Chem. Chem. Phys.* **2014**, *16*, 26730-26734.

19

Accelerators

❑ **3rd generation : Syncrotron, construction cost 140M USD**

❑ **4th generation : XFEL, construction cost 400M USD.**

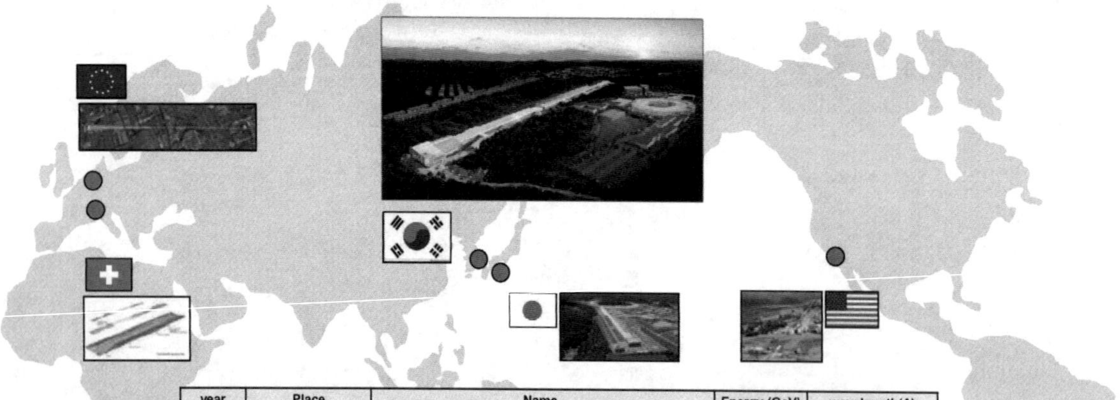

year	Place	Name	Energy (GeV)	wavelength(A)
2009	Menlo Park, CA, U.S.A.	LCLS (LINIAC Coherent Light Source)	14.3	1.5
2011	Harima, Japan	SACLA (Spring-8 Angstrom Compact free-electron Laser)	8.0	0.8
2015	Pohang, Korea	PAL-XFEL (Pohang Light Source - X-ray Free Electron Laser)	10.0	1.0
2015	Hamburg, Germany	European XFEL	17.5	0.5
2016	Villigen, Switzerland	SwissFEL	5.8	1.0

Liquid water structure

❖ Solute-induced perturbation of water structure

- The dissolution of a solute leads to the perturbation of the structure of liquid water.

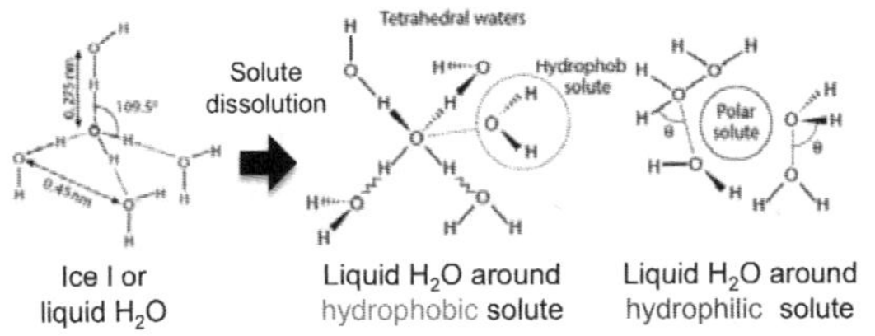

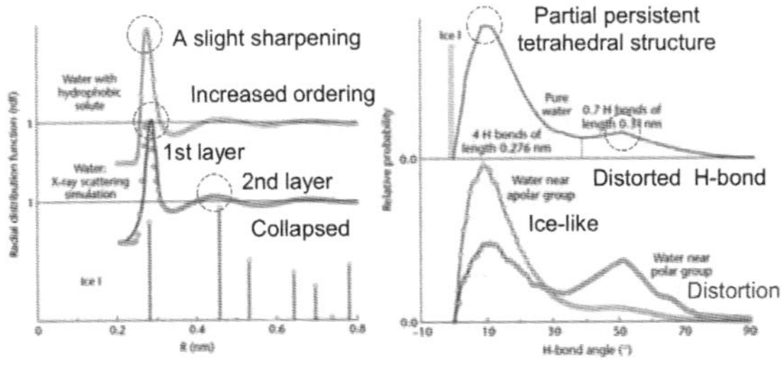

❖ Adsorption vs. Perturbation

- Kinetic hydrate inhibition was postulated to occur via a combination of two different mechanisms.

"Adsorption inhibition"	"Perturbation inhibition"
- Adsorption onto the hydrate surface or nucleating sites	- Disruption of the water structure
- Direct contact with surface	- Increase in barrier to nucleation
	- Decreased ordering

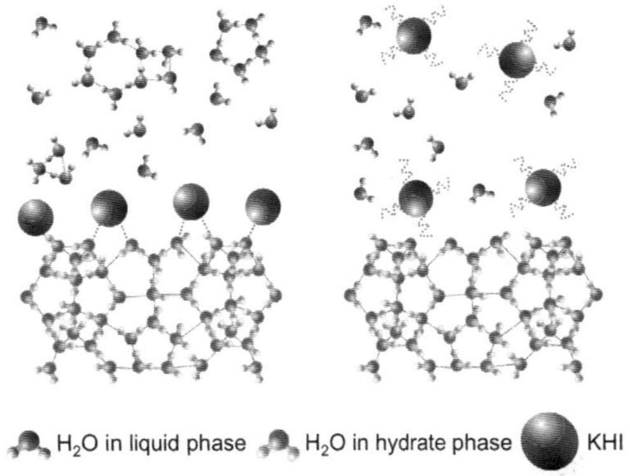

E. D. Sloan, C. A. Koh, *Clathrate Hydrates of Natural Gases*, 3re ed., CRC Press, FL, 2008.
Water: Structure and Properties, *Encyclopedia of Life Sciences*, 2001.

21

Water structure perturbation

❖ Polarized Raman spectroscopy

- Collective band: In-phase stretching motion of strongly hydrogen bonded water
- L-ala, L-asp, L-asn, and L-his disrupted the water structure, whereas L-phe strengthened.
- The extent of perturbation correlated well with the hydrophobicity.

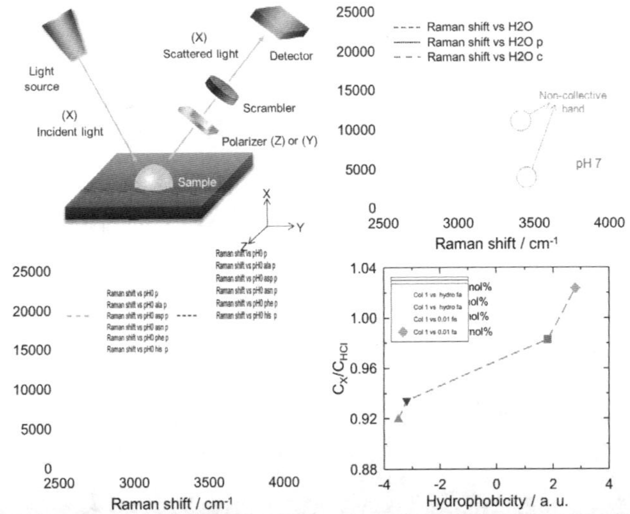

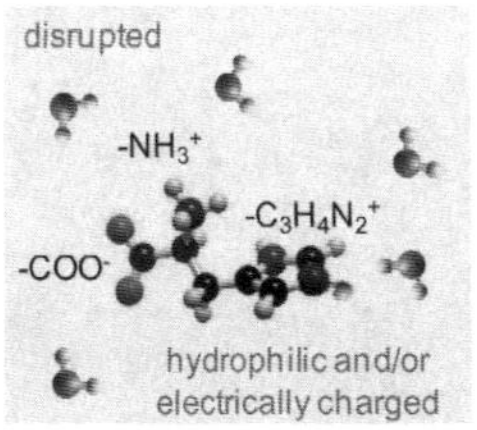

Water structure around L-his

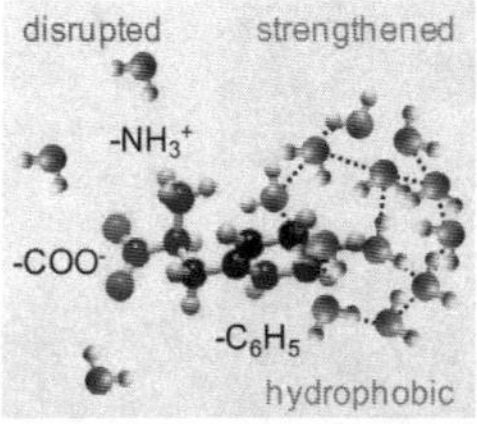

Water structure around L-phe

Sa et al. *Sci. Rep.* **2015**, *5*, 11526.

22

KHI mechanism

❖ Perturbation inhibition

- Amino acids with lower hydrophobicity were found to be better KHIs, working by disrupting the water hydrogen bond network, while those with higher hydrophobicity strengthened the local water structure.

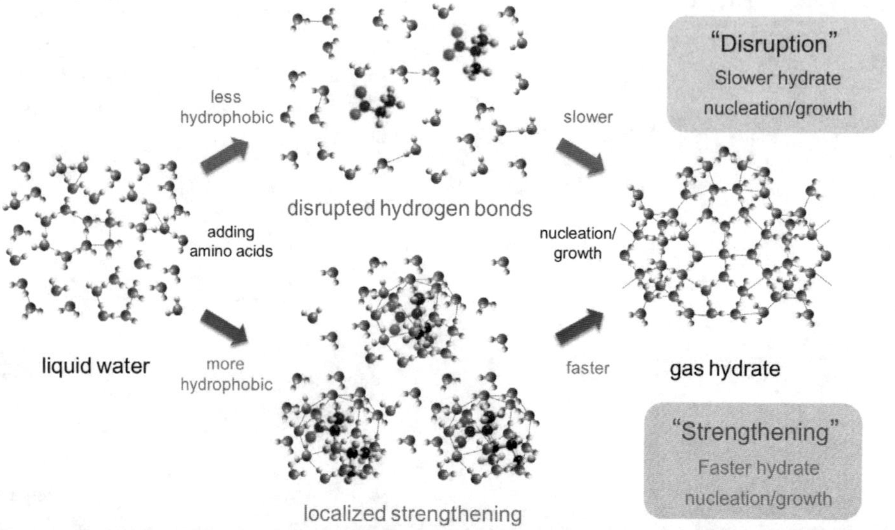

Sa et al. *Proceedings of the 8th International Conference on Gas Hydrates (ICGH 2014)*, Beijing, China, July 28-Aug 1, 2014. 23

In Honor of Stanley Sandler – (74e)

Summary & thanks

1. QC method

- Resonance frequency & Resistance change with phase transition
- Continuous measurement (Memory effect)
- Quick screening of hydrate promoters (30 times faster)

2. Amino acids

- Promising inhibitors: -COOH & $-NH_3$ (Hydrogen bond), Zwitterions (Electrostatic interaction)
- Dual Function : Thermodynamic inhibition & Kinetic inhibition
- Inhibition mechanism : Perturbation of hydrogen bond network of liquid water
  Depress nucleation

I hope to participate in a session "In honor of Stan's 80th Birthday" with thermodynamics.

24

414461 Gas Hydrate Phase Equilibria at Extreme Conditions

Abstract:

Gas hydrate phase equilibrium and hydrate formation kinetics in the presence of high salt concentration (near saturation in solution) is investigated to study the interplay of hydrate formation and scale deposition. There are selected data for hydrate phase equilibrium with high salinity systems, but there are no data for these systems at ultra-high pressures (> 15,000 psia). These data are increasingly important and offshore oil production moves to more challenging production environments. This study is focused on the development of the equipment and measurements of gas hydrate phase equilibrium at extreme conditions, that is, high salt concentrations and ultra high pressures. It is also necessary to verify the existing prediction models for hydrate phase equilibrium in high salinity systems, and assess whether improvements must be considered in the models incorporated for each of the available tools, such as CSMGem, Multiflash®, PVTSim®, DHR Hydrate®, and HydraFLASH®.

411285 Thermodynamics in CO2 Capture By Chemical Absorption

Abstract:

One of the most important CO2 capture methods is the chemical absorption. Amine-based solutions have been employed as absorbents to capture CO2. Common shortcomings of these absorbents include high regeneration energy, toxicity and corrosive nature. In this paper, we shall analysis the needs of new solvent from a thermodynamic view point. Thermal energy required for regeneration was broken down to desorption heat, sensible heat, and latent heat of solvent loss. A quick screening experiments were introduced for testing new absorbent candidates by measuring absorption rate and capacity, heat capacity, desorption energy and solvent loss. Aqueous solutions of a series of tetra-alkyl ammonium amino acid salts were evaluated and benchmarked against 30wt% monoethanolamine (MEA) solution. They were selected because of their ability to reduce vapor pressure of water. The results show that [N2222][L-Alanine] and [N1444][L-Alanine] both possess high absorption and desorption efficiency without precipitation problem. The study illustrates the importance of thermodynamics consideration in guidng process development and how critical thermodynamic data can be used for benchmarking process alternatives.

Representation and Validation of Liquid Densities for Pure Compounds and Mixtures

John O'Connell[1]*, Vladimir Diky[2], Jens Abildskov[3], Kenneth Kroenlein[2], Michael Frenkel[2]

[1] Professor Emeritus, University of Virginia, Nipomo, CA
[2] National Institute of Standards and Technology, Boulder, CO
[3]CAPEC-PROCESS, Technical University of Denmark, Lyngby, DK

Honoring Stanley Sandler's 75th Birthday

Thermophysical Property Data

- Amount doubles each decade
- ~30% papers contain errors in values, metadata, & uncertainties
- Errors propagate & cause damage in analyses & models
- Data validation essential in critical evaluation to:
  - Identify questionable data
  - Prevent erroneous use
  - Increase "value" of
    - Information about accuracy & reliability
    - Knowledge to enforce consistency among properties
    - Wisdom to reliably support applications
- NIST ThermoData Engine (TDE) implements methods

Liquid Volumetric Data

- No rigorous consistency test (*e.g.*, Gibbs-Duhem Equation used for partial molar properties)
- Validation currently involves finding:
  - Outliers via
    - Examining trends (e.g., density vs. pressure)
    - Comparing independent measurements
  - Systematic issues via models
    - Compressed liquids: e.g.,Tait model or multiparameter EOS
    - Unsatisfactory if no theoretical basis or too many parameters
- Present work describes successful use of 3-parameter Corresponding States Model for
  - Densities > 1.5 $\rho_{critical}$
  - Temperatures < $T_{critical}$

Correlation of Reduced Bulk Modulus

- Reduced Bulk Modulus, *B*, related to integral, *C*, of molecular direct correlation function, *c*, (DCFI)

$$B \equiv \left.\frac{\partial P/RT}{\partial \rho}\right|_{T,\mathbf{x}} = \frac{1}{\rho\kappa_T RT} = 1-\rho\int c(r;T,\rho)d\mathbf{r} \equiv 1-C$$

- Corresponding States Correlation for *C* (Huang, 1987)

$$1-\left.\frac{\delta(P/RT)}{\delta\rho}\right|_T = 2-C(T,\rho)$$

$$= C^*\left\{\sum_{i=0}^{3}\sum_{j=0}^{2} a_{ij}(\rho/\rho^*)^i (T/T^*)^j\right\}$$

- a_{ij} universal, insensitive T/T^*, isotherm crossover at C^*
- fitted C^*, ρ^*, T^* to over 300 substances & mixtures

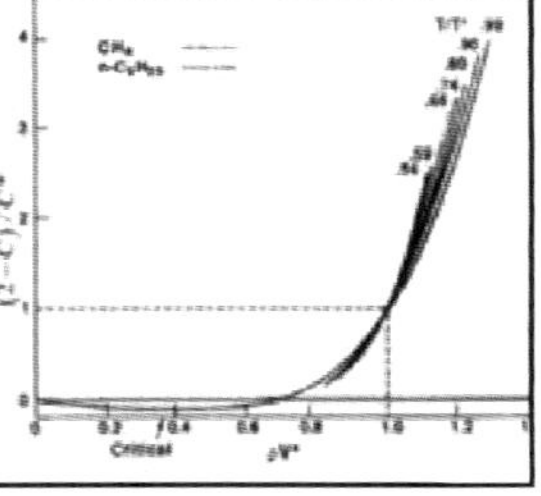

Pressure-Density Model

- Pressure-density relation from integrating B

$$\frac{P-P_0}{RT}=\int_{\rho_0}^{\rho}\sum_i\sum_j x_i x_j\left[1-C_{ij}\left(T,\rho,\{x\}\right)\right]d\rho$$

- Given T, P, $\mathbf{x}$, find ρ from standard state of P^0, ρ^0 for pure or mixed liquid at low P^0
- Most reliable current model

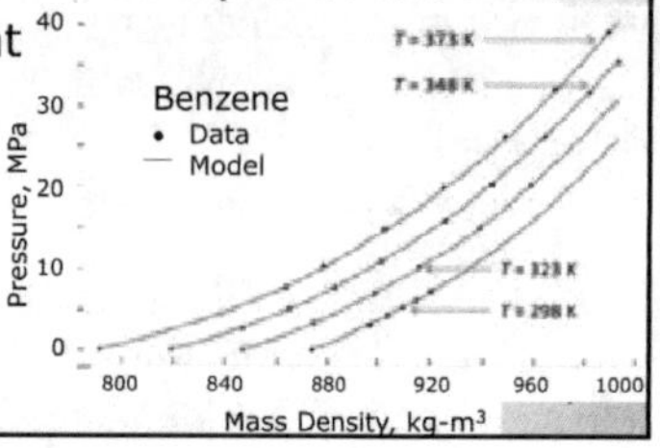

Examples I. Accurate Data

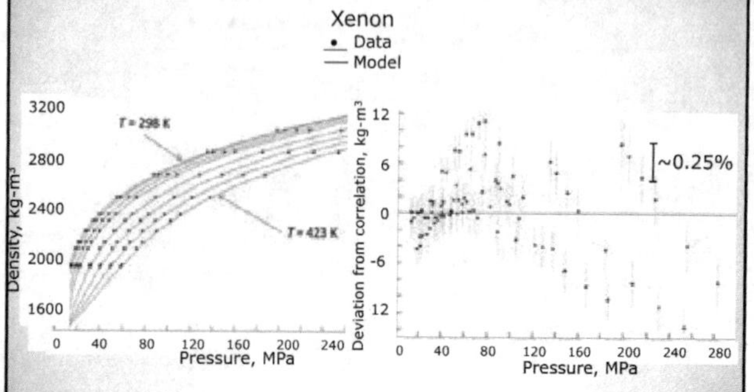

Examples II. Erroneous Data

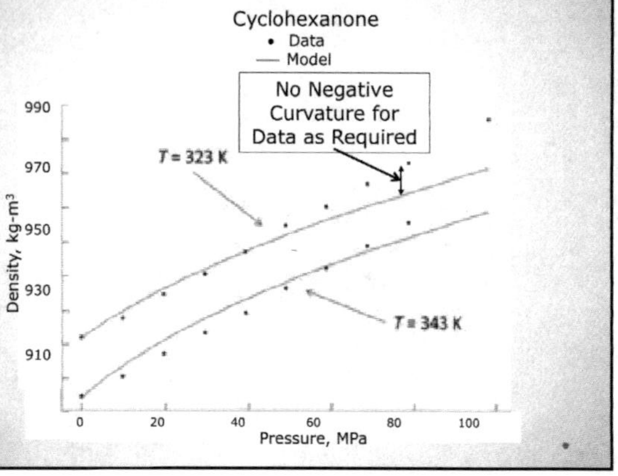

Examples IIIa. Publishing Typos

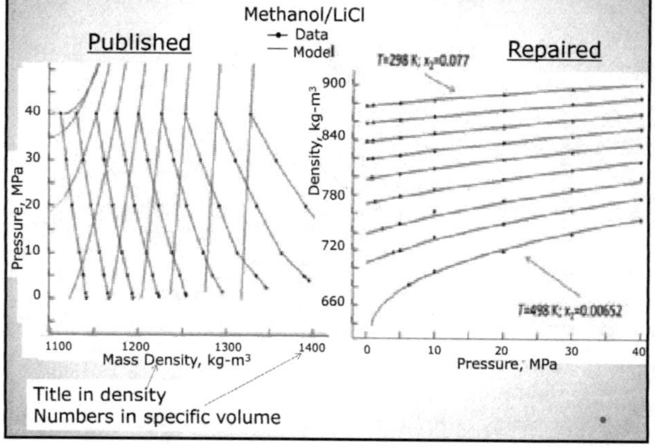

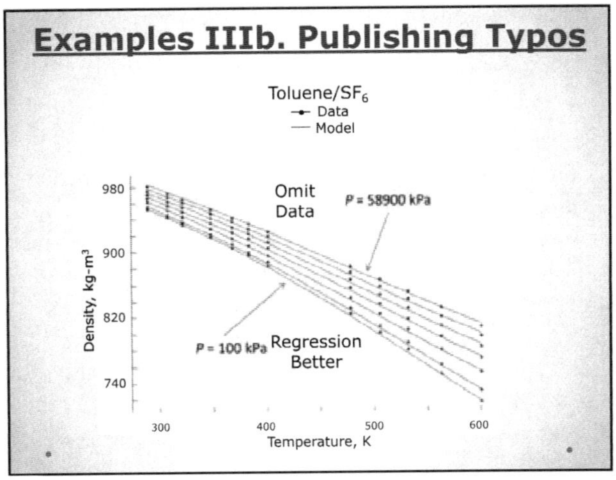

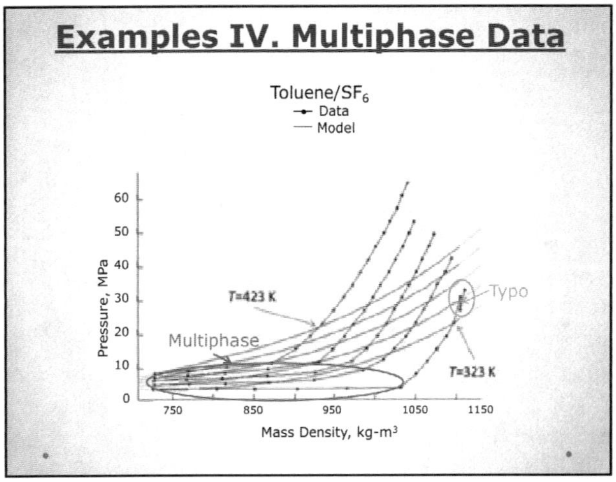

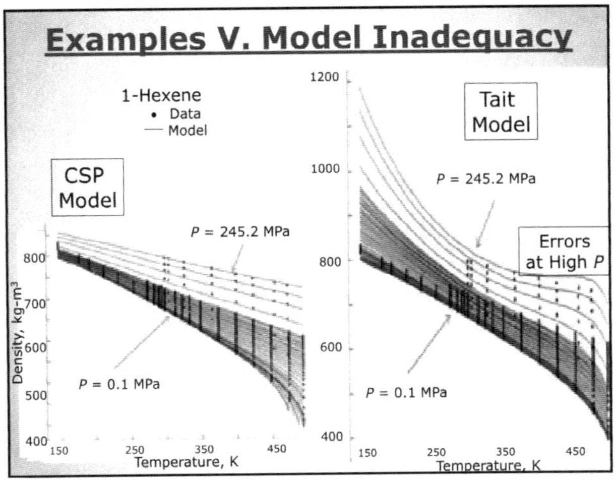

Conclusions

- Liquid density data validation established
  - Based on CSP formulation of reduced bulk modulus
  - Allows uncovering of variety of errors
    - Random & systematic measurement errors
    - Textural & typographical reporting errors
- Implemented in NIST TDE software (trc.nist.gov/tde.html)
- References:
  - Original CSP model – Huang & O'Connell *Fluid Phase Equil.* **1987**, 37, 75-84
  - Applied to ionic liquids - Abildskov, et al. *Fluid Phase Equil.* **2010**, 295, 215-229
  - TDE implementation - Diky, et al. *J. Chem. Eng. Data* ASAP **2015**, DOI: 10.1021/acs.jced.5b00477

435185 Stanley Sandler's Contributions to the Development of Equations of State

Abstract:

Professor Sandler has authored or co-authored nearly 400 refereed journal articles, and approximately 20 percent of these papers are directly related to the development of equation of state theory including many publications on novel mixing rules that combine excess free energy (activity coefficient) models with popular cubic EOS's (Peng-Robinson, Soave-Redlich-Kwong, van der Waals EOS's). Other notable work includes the introduction of a generalized van der Waals partition function as a basis for EOS mixing rules and with application to square-well fluids. Moreover, Professor Sandler's group applied Wertheim's first-order perturbation theory for the dimer reference fluid to obtain equations of state for hard-sphere chain fluids, and they obtained useful new molecular simulation data for the hard-dimer radial distribution function at contact value. Dr. Sandler and one of his students also developed an accurate and greatly simplified dispersion term for the statistical associating fluid theory (SAFT) EOS's. He has made several other contributions in this area, which include several book chapters and a few complete books dedicated to equations of state.

This talk will highlight specific milestones in the career of Professor Stanley I Sandler specifically related to the development of equations of state, and the author will make connections to his own work where the honoree has provided inspiration and motivation.

412571 Sodium Octa-Vacancy Patterns in O3-NaMnO2 during Cycling

Abstract:

Sodium octa-vacancy patterns in O3-NaMnO2 during cycling

Masoud Aryanpour and Young-Gyoon Ryu

Advanced Materials Lab, SAIT-America

255 Main Street, Cambridge, MA-02142

Transition metal oxides have been under extensive experimental research to work as electrode materials in Na-ion batteries [1, 2]. Na-based electrochemical cells are among viable candidates to reduce the price of advanced batteries through offering similar, if not better, features provided by Li-ion chemistries.

Mere replacement of Li with Na in the known Li-TM-O compounds does necessarily lead to optimized materials for Na chemistry. Both theoretical and experimental results point to crucial differences between the two classes stemmed from the nature of their alkaline ion component. Despite the larger ionic radius of Na^+, the ionic diffusion of Na in some layered oxide structures is even better than that of Li^+.

As these cathode materials are charged, Na atoms are de-intercalated from the host structure, leaving behind their empty crystallographic sites. Vacant sites in the sodium layers create a driving force for the other ions to adjust their initial position, which promotes various phase changes. It is thus important to understand the effects of vacancy formation in the Na layers of electrode materials.

An important class of oxide materials is based on the layered O3-NaMnO2 [3,4], whether purely made of Mn or mixed with other transition metals such as Fe, V, Cr, Co, and Ni. It has been shown recently that the intra-layer patterns of Na atoms in this class of materials strongly depend on the chemistry of the involved transition metal [5].

In this computational work, we study the in-layer patterns formed by Na-vacancies in Na_xMnO2 at x=0.5. This Na concentration corresponds to the half charging state, where superstructures might form as a result of provided empty sites. Figure 1 shows one of the Na patterns that may form in this material. Small spheres represent vacant sites, while the large ones stand for an occupied Na site.

[1] BL Ellis and LF Nazar. "Sodium and sodium-ion energy storage batteries", Current Opinion in Solid State and Materials Science, 16(4):168-177, 2012.

[2] C. Didier, M. Guignard, C. Denage, O. Szajwaj, S. Ito, I. Saadoune, J. Darriet, and C. Delmas, "Electrochemical Na-Deintercalation from NaVO2", Electrochemical Solid-State Letters, 14(5), A75-A78, 2011

[3] M.M. Doeff, M.Y. Peng, Y. Ma, and L.C. De Jonghe, "Orthorhombic NaxMnO2 as Cathode Material for Secondary Sodium and Lithium Polymer Batteries", Journal of Electrochemical Society, 141 (11), L145-L147

[4] X. Lin, et al, "Direct Visualization of the Jahn-Teller effect coupled to Na ordering in Na5/8MnO2, Nature Materials, 2014, DOI: 10.1038/NMAT3964

[5] M. Aryanpour, L. Miara, Y-G. Ryu, "Staging and In-Plane Superstructures Formed in Layered $NaMO_2$ {M = Sc, Ti, V, Cr, Mn} during Na De-Intercalation", Journal of Electrochemical Society, 162 (4), A511-A519, 2015

Figure 1 One possible configuration in the sodium layer of O3-Na_xMnO2 (x=0.5) when used as the cathode of Na-ion batteries. The computational unit cell contains 16 Na-sites per layer, which is only half-occupied.

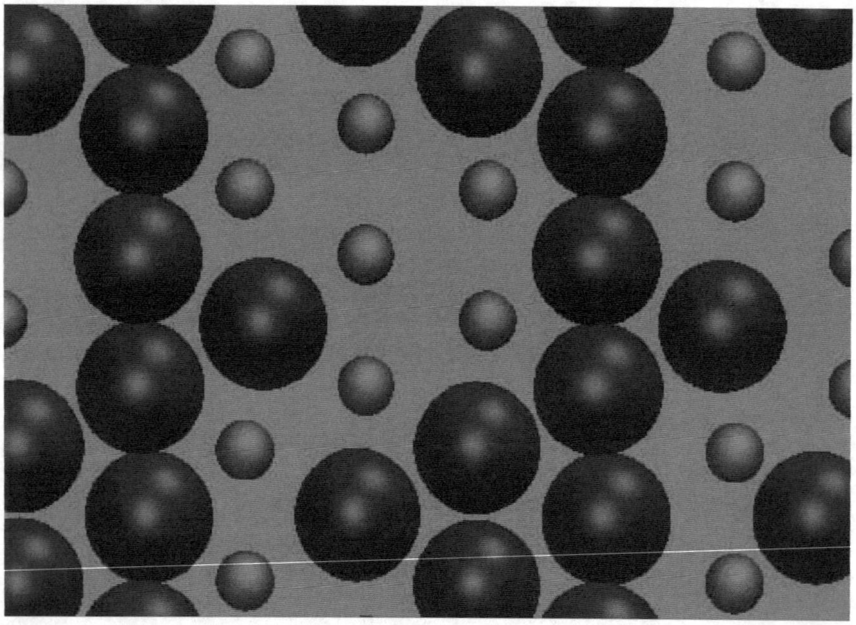

427340 The Crystal Structure of Coalescing Au and Its Alloy Nanoparticles By Ab Initio Molecular Dynamics

Abstract:

Abstract

Gold nanoparticles are used in a range of applications including electronics, catalysis, sensors, plasmonic biosensing and target-specific drug delivery as therapeutic agents. Physical properties (e.g. reactivity and ligand adsorption of gold nanorods) of such nanostructures depend on the particles crystal structure and surface facet orientation. Even though particle characteristics such as size distribution, primary particle and agglomerate size and morphology have been investigated extensively (Goudeli et al., 2015) little is known about particle crystallinity in atomistic scale. Experimental investigation of particles' nanocrystallinity with electron microscopy leads to incomplete characterization as it is based on the analysis of 2D projections of nanoparticles. However, Molecular Dynamics (MD) simulations can be used to provide physical insight in crystal structure changes and dynamics with atomistic detail (Buesser and Pratsinis, 2015) and can complement the above experiments, especially for very small nanoparticles.

Here, *ab initio*MD simulations are used to systematically investigate the sintering mechanism and crystallinity dynamics of gold and its alloy nanoparticles of various size and temperatures. The stage of crystallinity is theoretically investigated by the deviation of each gold atom from a perfect face cubic centered crystal. This deviation is quantified by the so-called bond order parameters (Steinhardt et al, 1983) which are measures of the local and extended orientational symmetries of the particle. The local degree of distortion of coalescing Au particles is described by the disorder variable (Kawasaki and Onuki, 2011). During adhesion, particles reveal increased degree of distortion compared to later stages of sintering, regardless of particle size and sintering temperature, while they form grains of different size and orientation. Large particles (e.g. 4 nm in diameter) form twin boundaries, consistent with experiments of Au nanoparticles coalescing by electron beam irradiation (Yuk et al., 2013). At high temperatures and sufficiently long time these grains transform to a single crystal.

The grain size and crystal direction depend on the synthesis process conditions (e.g. temperature) and lead to the exposure of different facets affecting surface chemistry and product particle morphology and their catalytic properties (e.g. catalytic activity and selectivity) which can be tuned by controlling particle structure and nanocrystallinity. Defects such as grain boundaries at the atomic scale can strongly affect electrical, optical and mechanical properties of 2D materials.

References

Buesser, B., and Pratsinis, S.E. (2015) *J. Phys. Chem. C*, 119, 10116-10122.

Goudeli, E., Eggersdorfer, M.L., and Pratsinis, S.E. (2015) *Langmuir*, 31, 1320-1327.

Kawasaki, T., and Onuki, A. (2011) *J. Chem. Phys.*135, 174109-1-174109-8.

Steinhardt, P.J., Nelson, D.R., Ronchetti, M. (1983) *Phys. Rev. B*, 28, 784-805.

Yuk, J.M., Jeong, M., Kim, S.Y., Seo, H. K., Kim, J., Lee, J.Y. (2013) *Chem. Commun.* 49, 11479-11481.

432538 Computational Screening of Metal-Ligand Systems for Methanol Carbonylation

Abstract:

Acetic acid is produced industrially by methanol carbonylation with iridium or rhodium based catalysts. The expense of iridium, and rhodium has led to attempts to improve catalytic activity by designing ligands for iridium and rhodium metal complexes.[1] Alternatively, the cost of the catalyst can be reduced, by designing ligands to improve the catalytic activity of a less expensive metal center.[2] Unfortunately, the methyl iodide promoter necessary for the reaction can often preferentially alkylate the ligands bound to the metal center. As a result, carefully designed catalysts often degrade to their unligated form in the reactor. Identification of metal/ligand pairs that are resistant to alkylation by methyl iodide as well as improve catalytic activity is necessary for a commercially viable catalyst.

Density functional theory (DFT) is used to screen ligands-metal pairs for stability and reactivity for methanol carbonylation. Metal centers with cobalt, rhodium, and nickel were examined with nitrogen, phosphorous, arsenic, and antimony based ligands. The bond dissociation energy of metal-ligand bonds are compared to ligand bonds in the undesired ligand salts that result from ligand alkylation by methyl iodide. Ligands that bond more strongly to the metal than in the salt are more likely to result in a stable ligated catalyst. The relative performance of metal-ligand pairs is also estimated by computing the activation barrier for the rate-limiting step of methyl iodide addition in the catalytic cycle. Our results are validated for some systems by existing experimental results, and suggest new metal-ligand combinations for further study.

References:

1. Thomas CM, Suss-Fink G. Ligand effects in the rhodium-catalyzed carbonylation of methanol. Coord. Chem. Rev. 2003; 243:125-142.

2. Moser WR, Marshik-Guerts BJ, and Okrasinski SJ. The mechasnism of the phosphine-modified nickel-catalyzed acetic acid process. J. Mol. Catal. A: Chem. 1999; 143:71-83.

434274 Liquid-Phase Mechanism Generation for Predicting Fuel Oxidation

Abstract:

We have added functionality to automatic reaction mechanism generation software, RMG[1], to more accurately predict the reaction rates of detailed, liquid-phase chemical mechanisms. One industrial application of liquid-phase mechanism generation, explored here, is in developing models for fuel oxidation. Other technical applications for this software include the upgrading of heavy oils with supercritical water, and the processing of microalgae to produce fuels.

Mechanistic similarities exist between radical reactions in the gas phase and liquid phase, but the presence of a solvent changes both the thermodynamics and reactivity of each chemical species in a kinetic model. For systems which may include hundreds or thousands of reactions, we use automatic mechanism generators such as RMG, which was first built for the gas-phase, to generate models faster and minimize errors. Estimates for thermodynamic and diffusion corrections to the gas-phase values have been added to RMG[2], but so far no systematic corrections to the reaction rates exist for when there are no experimental data available.

For a set of hydrogen abstraction reactions, M06-2X/MG3S calculations were used to find the geometries of reactants and transition states in both gas and liquid phase. A continuum solvation model, SMD, was used to find the energy of the liquid-phase structures in eight different solvents. Using these energies, the difference in barrier height between gas and liquid phase, ΔE, was computed; a positive ΔE corresponds to a higher barrier in solvent than in gas. Trends in ΔE based on reactant molecular structure have been deduced, and these trends hold across various solvents. Specifically, these trends include: (1) In hydrogen abstraction from alcohols, ΔE increases with proximity of abstraction site to --OH group; (2) Hydrogen abstraction from molecules with increasing number of carbons beyond C2 does not change ΔE; (3) ΔE is greater in abstraction from aromatic rings than the corresponding saturated ring.

We have implemented molecular group contributions to ΔE into a hierarchical tree in RMG in order to estimate this difference in barrier for reactions without data. Using the new data, rates of hydrogen abstraction in a fuel oxidation model[3]were modified. Reactor simulations of the resulting model were compared with those of the previous model, which did not include liquid-phase kinetic corrections. For other reaction types, solvation kinetic trends can be similarly developed.

[1] Green, W.H., Allen, J.W. et al. RMG – Reaction Mechanism Generator, Python Version. http://rmg.mit.edu, 2013.

[2] Jalan, A. et al. An extensible framework for capturing solvent effects in computer generated kinetic models. *J. Phys. Chem. B*, 117: 2955-2970, 2013.

[3] Ben Amara, A. et al. Toward Predictive Modeling of Petroleum and Biobased Fuel Stability: Kinetics of Methyl Oleate/ n-Dodecane Autoxidation. *Energy & Fuels*, 27: 6125-6133, 2013.

materials design

Force Field Optimization for Metal Oxide Fillers and Calculation of Mechanical Properties of Epoxy-Based Composites

David Rigby, Clive M. Freeman, Paul W. Saxe & Benoit Leblanc

2015 AIChE Annual Meeting,
Salt Lake City, UT, November 9, 2015

Background and Previous Work

- Continuation of project aimed at using atomistic simulations to perform quantitative prediction of polymer composite material properties
- Summary of related earlier studies
  - Optimized parameters in class II ‘PCFF’ force field included in LAMMPS distribution to improve prediction of PVT and cohesive behavior (-> **‘PCFF+’**) for common thermoset functional groups (amine, hydroxyl, ether, sulfone etc.).
  - Conducted basic studies on thermosets with different architectures, including:
    - Gel points and mechanical properties (see, for example "Advanced Composites for Aerospace, Marine, and Land Applications", T. Sano, T.S. Srivatsan and M.W. Paretti, eds., Ch14, pp 157-171, J. Wiley 2014)
  - Performed preliminary studies on silica-epoxy layered systems with and without added primer (AIChE presentation, Atlanta 2014).
    - Identified significant limitations associated with silica force field parameters within the original force field…

This Work: Goals

- Perform reparameterization of SiO_2 force field parameters, based on least-squares fitting of DFT-based QM data.
- Validate SiO_2 FF with reference to properties of interest (mechanical, bulk density etc.) over a range of temperatures.
- Create selection of models of layered quartz-epoxy systems, with and without added primer.
- Examine small strain elastic constant behavior, and investigate feasibility of quantifying effects of primer surface coverage.

SiO_2 Parameter Refinement

Silica Forcefield

- PCFF, based on the class II functional form of Hagler and coworkers (*), essentially represents SiO_2 as a covalent structure, with explicit terms describing bond stretching, angle bending, torsion, etc, as used routinely for organic species (advantage: ease of handling hydroxyl and other species chemically bonded to surface; disadvantage: melting, etc. not readily accommodated).

- The parameters used in the original PCFF closely resemble those published by Hill & Sauer (J. Phys. Chem. **98**, 1238 (1994)), which were derived by fitting to *ab initio* data calculated for small fragments of silica and zeolite structures.

- For SiO_2 and a variety of zeolites, the original force field predicts minimum energy structural features such as bond lengths and angles quite well; however temperature dependent properties and mechanical behavior are not so well-reproduced.

- see, e.g., J. Comp. Chem. **15**, 162 (1994); JACS **116**, 2515 (1994)

Silica Forcefield Refinement

Since *ab initio* structure predictions of materials such as silica, zeolites and other crystallisable compounds can now be performed using **periodic** quantum mechanical techniques, it is feasible to use such data directly, avoiding the need to study isolated clusters, fragments or small molecule anaologs when performing parameterization.

In this work, we used the VASP simulation package to generate the raw data – energies, forces, stresses – required for fitting the class II forcefield parameters for SiO_2.

Typically, the QM data were generated by performing molecular dynamics simulations at elevated temperatures to sample a representative region of the energy surface, with snapshots saved at regular intervals.

Parameterization: First-Principles DFT Target Data

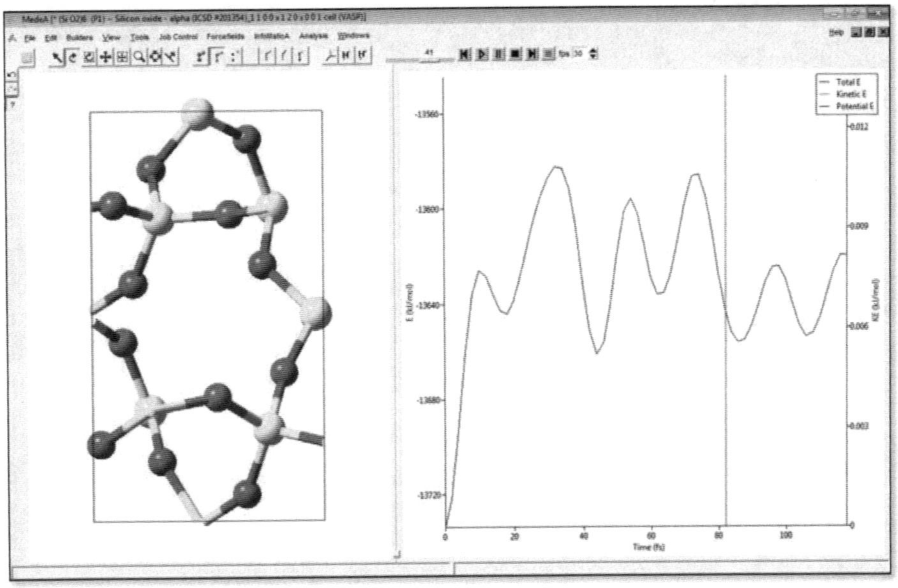

- Used α-quartz structure from ICSD database (ICSD 201354)
- VASP MD: PBE, 400eV cut-off, k-spacing 0.5Å^{-1}, 120fs, 2fs time step, 1000K

DEV
30732

Setting Fitting Targets

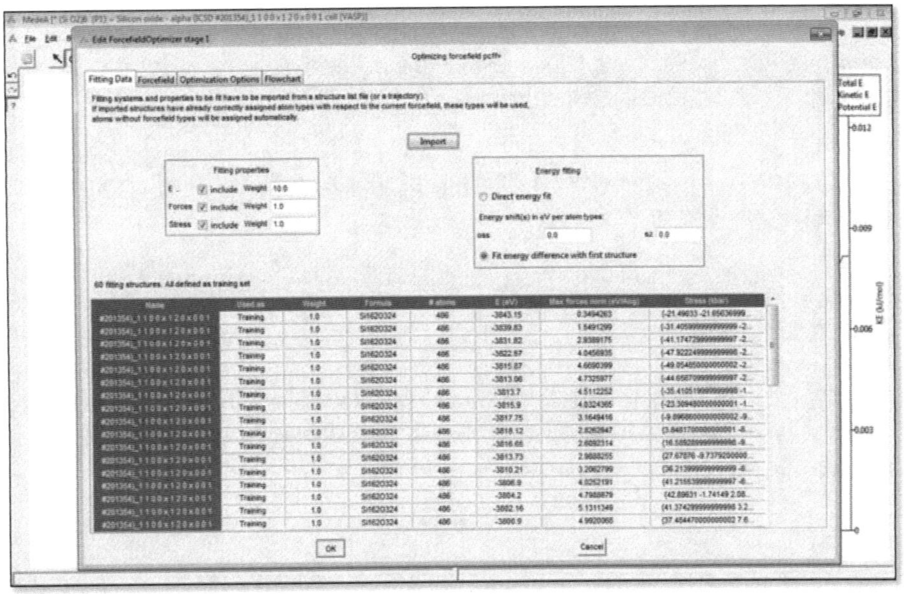

- Energy, forces, and stresses used in fitting
- Weight on energy increased by factor of 10 relative to forces and stresses

Parameters Optimized

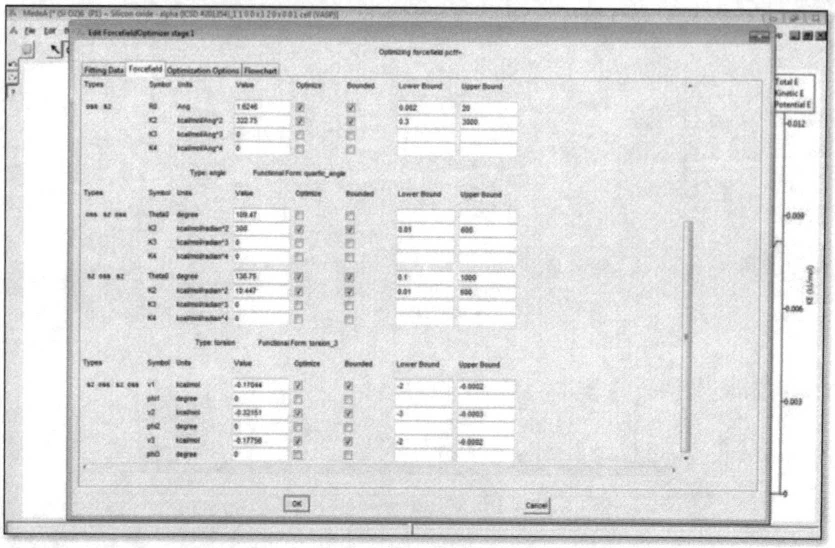

- Cross terms set to zero (not shown here)
- Initial parameterization: Quadratic form imposed for bond and angle terms
- Using Levenberg-Marquardt least squares optimization

Notes on Optimization Process

- Preliminary investigation showed that using the original (partial) atomic charges on Si and O significantly, and adversely, affects the calculated density.

- Reducing oxygen charge to -0.16 on oxygen atoms and 0.32 on silicon atoms reduces density discrepancy to 6.4%

.

- Small strain elastic properties significantly improved versus initial parameters.

- Finally, adjusting r_0 for Si-O using the experimental crystal data allows a good match with density to be achieved over a broad temperature range, while preserving good agreement of elastic constants.

Final Forcefield Fitting Results

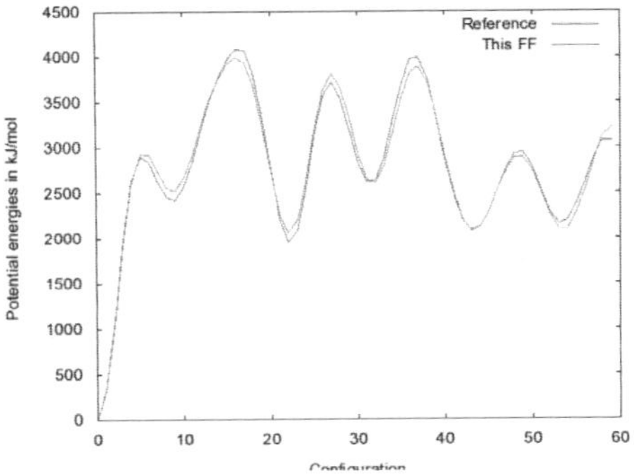

Agreement with target energies and forces is very good with revised charge model

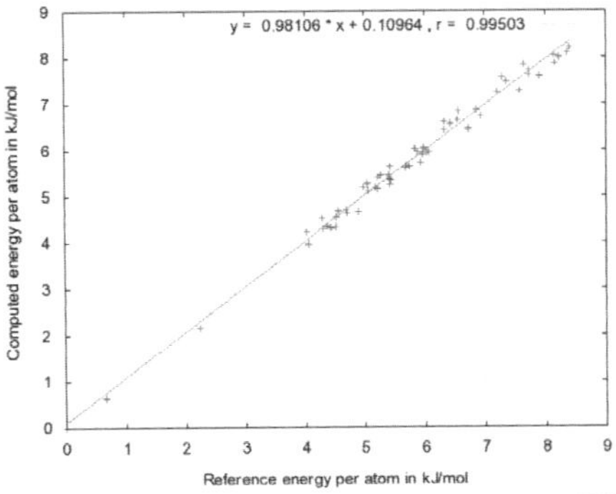

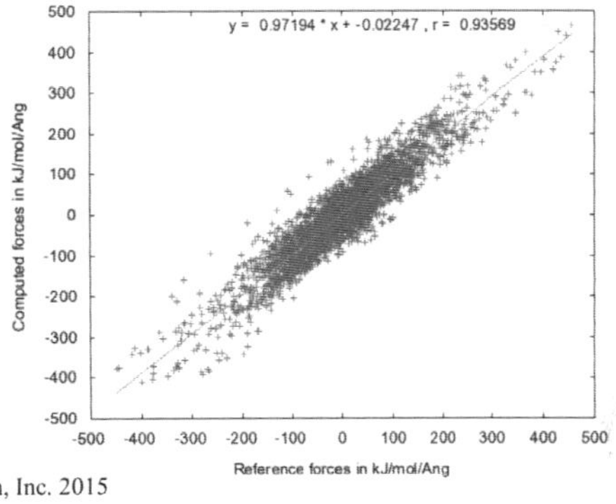

Computed 298K α-Quartz Properties

Method	298K Density (g/cm^3)	Bulk Modulus (GPa)	Young's Modulus (GPa)	Shear Modulus (GPa)
Original PCFF	2.45	97	203	88
PCFF+ updated	2.64	39	82	35
Expt.	2.65	36-38	73-96	31-41

- Forcefield optimization yields an improved model showing good agreement with first-principles results and experimental mechanical properties (mechanical properties were not included in fitting)

Temperature Dependence of α-Quartz Density

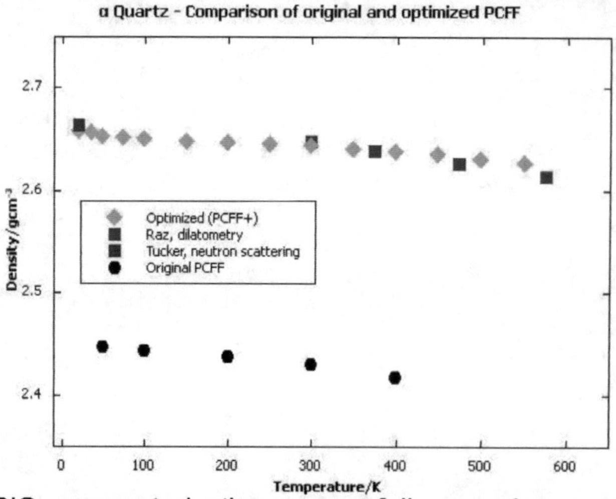

- Refined SiO_2 parameterization successfully reproduces quartz density over temperature range of interest in SiO_2-Thermoset composite studies (using LAMMPS NPT dynamics simulation).
- Slightly low CTE of calculated data at highest temperatures may be caused by neglect of cubic and quartic terms in fitting?

SiO_2-Thermoset Composite Studies

Earlier Work: Bulk Thermosets

- Experimental data for thermosets formed using crosslinkers with different architecture show a fairly significant reactant functionality effect on mechanical properties

- In earlier studies, we examined elastic constants of bulk resins with different architectures, cured with 4,4' diaminodiphenylsulfone (DDS) architecture;:

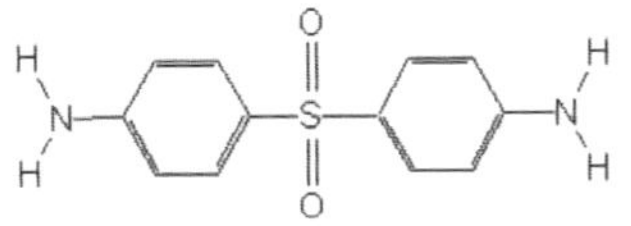

- Approximately 100 thermoset models were examined for resins of 3 types:
- 'Type RB2' : Diglycidyl ether of bisphenol A (DGEBA)
- 'Type RB3' : Triglycidyl p-amino phenol (TGAP)
- 'Type RB4' : Tetraglycidyl diaminodiphenylmethane (TGDDM)

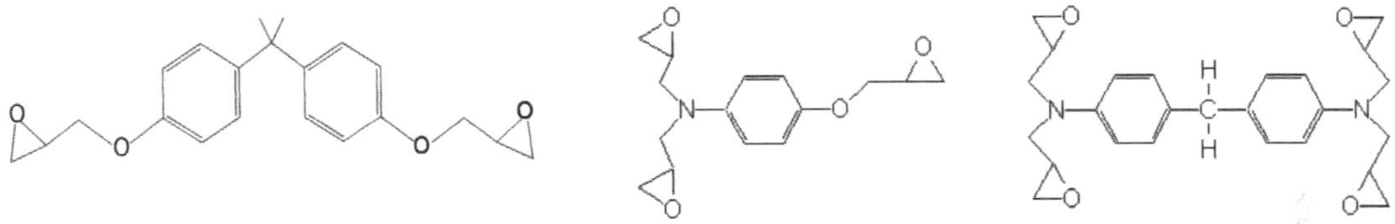

Bulk Thermosets – Mechanical Properties

- Application of Hill-Walpole analysis yields narrow bounds, and correctly predicts resin architecture effect on tensile moduli.

Type	Resin	Calculated Bounds (GPa)	Experiment (GPa)
RA_4+RB_2	**DGEBA**	3.49-3.53	2.4-3.2[a]
RA_4+RB_3	TGAP	4.42-4.45	4.396±.027[b]
RA_4+RB_4	TGDDM	5.18-5.19	5.103±.033[c]

a. Extents of reaction relatively broad **0.5-1.0**; ~300K; using dynamic mechanical analysis at 1Hz (White, 2002)
b. Extent of reaction 0.93; 295K; strain rate 1.67 x$10^{-2}s^{-1}$ (Behzadi, 2005)
c. Extent of reaction 0.88; 295K; strain rate 1.67 x10^{-2} s^{-1} (Behzadi, 2005)

Composite Studies – Force Fields

- Used PCFF+ refinement of PCFF
  - For SiO_2 – includes newly-reparameterized covalent network parameters.
  - For organic component (epoxy and primer) – also includes recently-improved parameters for secondary and tertiary amine groups found in many epoxies.

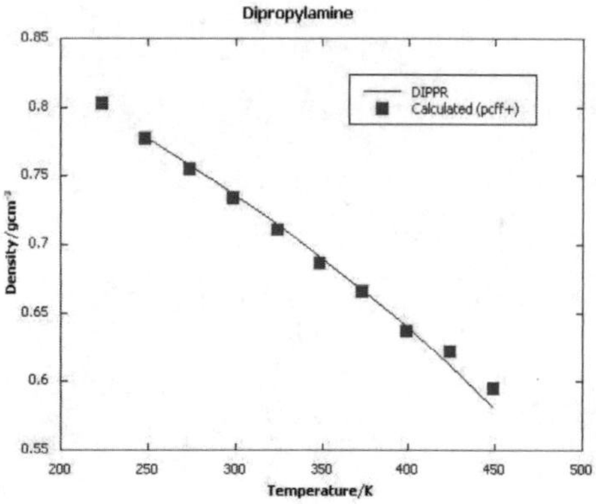

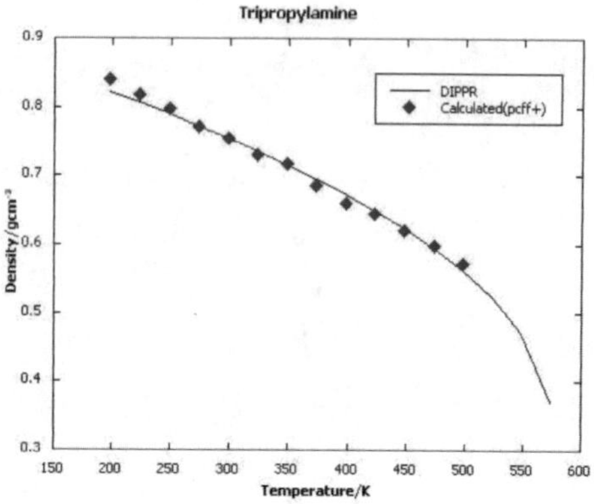

Epoxy-Oxide Interfacial System Studies

In general, interfacial composite systems of interest contain:

- Various epoxies (amine, anhydride-cured; epoxy-phenol-novolac, etc.)
- Oxides (**SiO_2**, Al_2O_3, TiO_2, etc.)
- Primers

Specifically, in this work, we focused on elastic constant calculation for systems containing:

- DGEBA epoxy crosslinked with 4,4' diaminodiphenylsulfone (as studied previously in bulk)

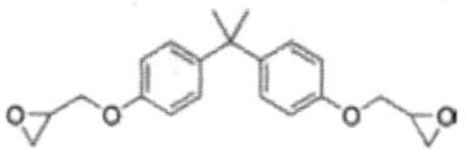

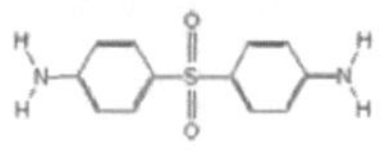

- SiO_2 (hydroxylated α-Quartz)

- γ-aminopropyltriethoxysilane

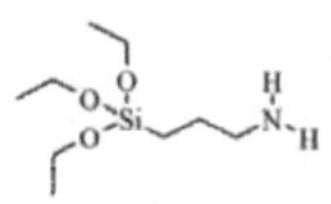

Epoxy-SiO_2 Interfaces – Model Building

Epoxy-coated primed silica interfacial systems created in a series of steps:

1. Preparation of isolated layers: silica – primer – unreacted epoxy
2. Silica, modeled as the dominant (100) surface of α-quartz - Rohl & Gay, Mineralogical Magazine (Dec 1995). All available surface sites hydroxylated, corresponding to **silanol number ~7.5/nm^2**
3. Covalent bonding of APS primer to silanol groups on silica surface
4. Equilibration of unreacted epoxy layer at primed SiO_2 surface
5. Simultaneous crosslinking and bonding of DGEBA epoxide groups to curing agent and APS amine groups
6. Equilibration of final interfacial system

(all covalent bonding operations – i.e. primer bonding and thermoset crosslinking - carried out using the MedeA® Thermoset Builder, with molecular dynamics equilibrations using the Sandia LAMMPS simulation program and PCFF+ force field)

In pictures…

Epoxy-SiO_2 Interfaces – Model Building

α Quartz (100) surface

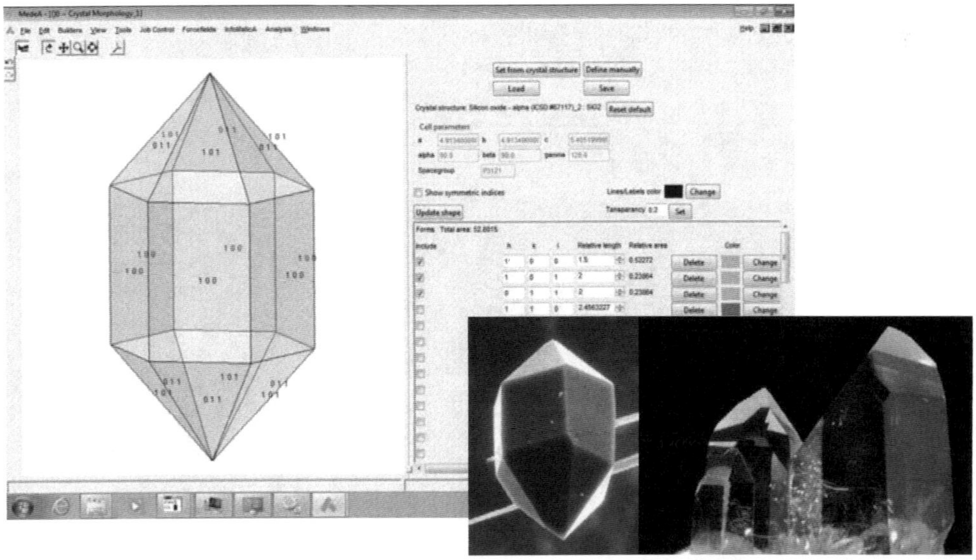

Epoxy-SiO_2 Interfaces – Model Building

α Quartz (100) + Primer + Unreacted epoxy

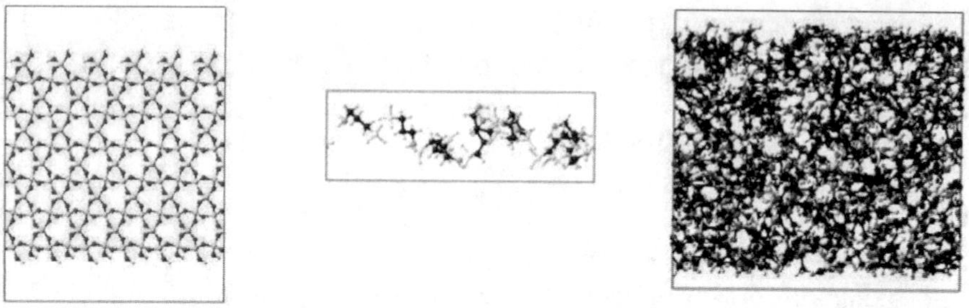

Epoxy-SiO_2 Interfaces – Model Building

γ–APS-coated (crystalline) silica (covalently bonded)

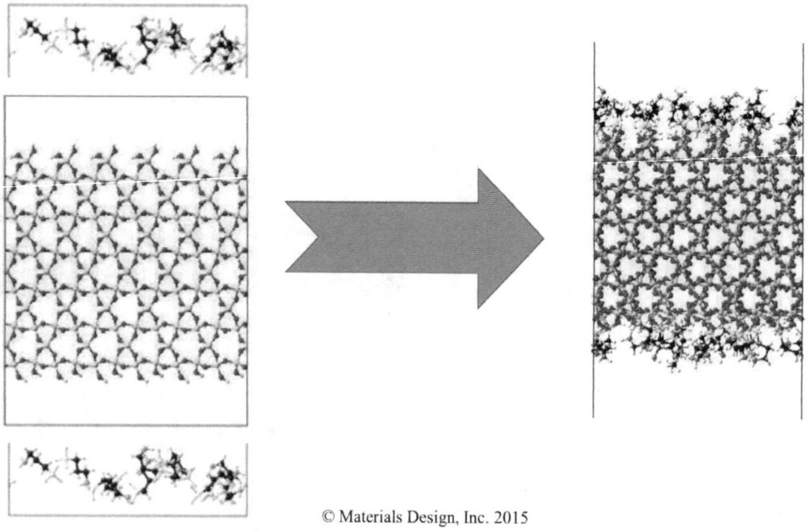

Epoxy-SiO_2 Interfaces – Model Building

Primed SiO_2+unreacted epoxy, crosslinking and equilibration

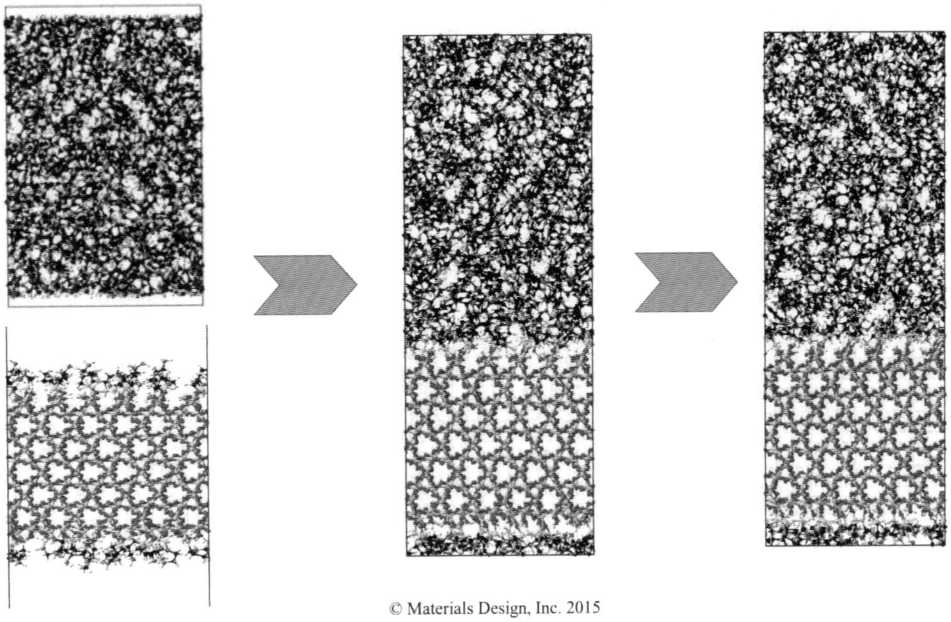

Epoxy-SiO_2 Interfaces – Mechanical Properties

Following equilibration, epoxy-SiO_2 systems were subjected to elastic constant calculations, as previously applied to bulk systems.

- Without primer
- Including primer with 12.5%, 25% and 50% of surface hydroxyls covalently bonded to primer, and exact stoichiometric balance of epoxide:amine groups present
- Compare C_{33} (perpendicular to interface) stiffness coefficients for pure epoxy and various silica-containing primed composites...

Epoxy-SiO_2 Interfaces – Mechanical Properties

Stiffness coefficient C_{33} for SiO_2 - epoxy composite layer systems with various levels of primer coverage:

Primed surface sites (%)	C_{33} (GPa)
0	7.2
12.5	7.8
25	8.8
50	5.0
Reference: Bulk Cured Resin	5.97 +/-0.29

Initial results suggest **intermediate** primer coverage produces optimal enhancement in small strain mechanical behavior.

Summary and Ongoing Work

- We continued previous polymer composite related studies on thermosets interfaced with inorganic oxide, focusing on:
  - SiO_2 forcefield parameter refinement
  - Model building approach, with additional force field parameterization
  - Elastic constants for bulk epoxy, unprimed SiO_2-epoxy, primed SiO_2-epoxy
  - (Also investigating heat transfer - in progress)
- Principal observations:
  - Structural and mechanical properties of SiO_2 are well-reproduced using the revised force field.
  - Indications are that maximum enhancement of stiffness coefficients occurs at intermediate levels of primer surface coverage

Summary and Ongoing Work

- Ongoing and planned future work:
  - Further SiO_2 parameter refinement with:
    - Expanded sets of QM training set data, using VASP MD simulations on additional SiO_2 polymorphs
    - Investigation of effects of using anharmonic bond and angle terms on high temperature ρ-T behavior (e.g. examine ability to reproduce accurately transition between α and β quartz)
  - Thorough assessment of sampling required to attain precision previously obtained with bulk thermoset properties (estimate 50-100 independent configurations required to adequately sample distribution of, e.g., C_{ij})
  - More detailed probing of optimal primer surface coverage
  - Computation of additional composite system properties:
    - Thermal conductivity
    - Large strain deformation/fracture behavior

Thank You!

Questions…

414746 Parameterizing Mosced with Molecular Simulation and Electronic Structure Calculations

Abstract:

Accurate and efficient models to predict the phase-behavior of nonelectrolyte solids in a wide range of solvents are central to the design of novel separation processes. A promising design tool is the MOSCED limiting activity coefficient model which is parameterized for 133 solvents. However, before predictions may be made for a solute of interest, solute MOSCED parameters are required. For novel compounds of interest, the necessary MOSCED parameters are unavailable and sufficient reference data is likely unavailable to regress the necessary parameters.

In this study we explore the use of molecular simulation and electronic structure methods to generate the reference data necessary to obtain solute MOSCED parameters. Specifically, we use solvation free energy calculations to predict the limiting activity coefficient of the solute in a limited range of solvents. This reference data is then used to regress the necessary MOSCED parameters for the solute. The method is applied to a range of pharmaceutical compounds and environmental pollutants. The limiting activity coefficients computed using MOSCED may subsequently be used to parameterize Wilson's equation and make equilibrium solubility predictions in pure and multicomponent solvents.

419521 Prediction of the Solubility of Ethylene in Polyethylene Using Molecular Simulations at the Operating Conditions of LDPE Polymerization Process Equipment

Abstract:

Prediction Of The Solubility Of Ethylene In Polyethylene Using Molecular Simulations At The Operating Conditions Of LDPE Polymerization Process Equipment

Sudheer Gondu*, Jhumpa Adhikari**

*Department of Chemical Engineering, IIT Bombay
(e-mail: *sudheer7g@ iitb.ac.in, ** adhikari@che.iitb.ac.in)

Abstract

The aim of the present work is to study the fluid phase behavior of ethylene and polyethylene mixtures by employing molecular simulations and theoretical methods such as equation of state (EOS) modeling (perturbed-chain statistical associating fluid theory, PC-SAFT, and SAFT) [1, 2]. Thermodynamic properties (such as vapor-liquid equilibria data) are necessary for the design and optimization of the separators, downstream of the polymerization reactor; in the high pressure process for low density polyethylene (LDPE) production. The vapor-liquid equilibria (VLE) data reported in literature at the operating conditions of the high pressure separator (HPS) and the low pressure separator (LPS) are limited. In this work, molecular simulations have been employed to generate the necessary data in conjunction with the Nath, Escobedo and de Pablo (NERD) force field [3, 4]. We have used the Hybrid Monte Carlo (HMC) and the Gibbs Ensemble Monte Carlo (GEMC-NPT) techniques [4] to determine the VLE of the ethylene-polyethylene mixtures. First, we have performed the simulations at the conditions reported by Nath et al. [4] and observed that our results are in reasonable agreement with the literature data [4]. This has been followed by molecular simulations at the operating conditions of the HPS and LPS. The solubility data from our simulations have also been compared with that estimated from traditionally employed EOS for polymer systems, viz., PC-SAFT and SAFT. It has been observed that ethylene content in the polymer rich phase increases with pressure at constant temperature at both the HPS and LPS operating conditions; this is consistent with the trend reported by Bogdanovic [5]. In the entire temperature and pressure range of the operating conditions of the separators, the SAFT estimates are higher than the corresponding values determined from our simulations; while PC-SAFT predictions show lower solubilites as compared to molecular simulation predictions. The VLE data estimated by all the three methods have been used in sizing the HPS and LPS. The volume of the HPS computed from molecular simulation generated data is lower than that calculated using either SAFT or PC-SAFT predictions. However, for the LPS, the volume calculated from molecular simulation data is significantly higher than that from SAFT and PC-SAFT.

References

[1] Gross J., and Sadowski, G., 2001, "Perturbed-Chain SAFT: An Equation of State Based on a Perturbation Theory for Chain Molecules", Ind. Eng. Chem. Res., 40, 1244-1260.

[2] Huang, S.H. and Radosz, M., 1991,"Equation of State for Small, Large, Polydisperse,and Associating Molecules: Extension to Fluid Mixtures", Ind. Eng. Chem. Res., 30,1994-2005.

[3] Nath, S.K., Escobedo, F.A., de Pablo, J.J. and Patramai, I., 1998," Simulation of Vapor-Liquid Equilibria for Alkane Mixtures", Ind. Eng. Chem. Res., 37, 3195-3202.

[4] Nath, S.K., Banaszak, B.J. and de Pablo, J. J. "Simulation of Ternary Mixtures of Ethylene, 1-Hexene, and Polyethylene", Macromolecules, 2001, 34, 7841-7848

[5] Bogdanovic, V., 1985," Industrial Aspects of Phase Separation In Ethylene-Polyethylene System at Pressures of 15-30 MPa", Ind. Eng. Chem. Process Des. Dev., 24, 576-581.

417656 Development of a Molecular Model of Crude Oil: Applications in Enhanced Oil Recovery

Abstract:

We have developed a multi-scaled coarse-grained molecular modelling framework to compute the optimal salinity at which a microemulsion forms in an oil-brine-surfactant system. The optimal salinity plays an important role in determining the surfactant mixture used in Alkali-Surfactant-Polymer floods for Enhanced Oil Recovery (EOR). The framework uses Dissipative Particle Dynamics (DPD) to study phenomena at the oil-water interface, which are sensitive to the chemical compositions of the crude, brine and surfactant cocktail.

The DPD-based coarse-grained formalism, called the Method of Moments, predicts trends in EOR behaviour. It addresses the question: Which way will a system swing if certain experimental handles are varied? The experimental handles primarily concern the choice of surfactant – chemical structure, topology – and nature of the oil. The utility of such a tool lies in its ability to scan through a list of molecules – e.g., surfactants, crude components – and rank them according to the system's sensitivity to their presence. At this stage, the crude oil is typically represented by a mixture of linear alkane (dodecane) and a surface-active component. Our library of such surface-active crude components contains 3 members at the moment: nonylphenol, butylcarbazole and naphthenic acid.

The molecular simulations were validated against phase behaviour experiments that studied the effect of surface-active crude components on the optimal salinities of dodecane-brine-surfactant systems. Phase behaviour tests were conducted separately for each crude component. We observed that the optimal microemulsion formed at a lower salinity when the crude component was either nonylphenol or naphthenic acid. The addition of butylcarbazole did not alter the optimal salinity at all. The optimal salinity decreased by ~57% in the presence of 2.04 mol% of nonylphenol and 2.45 mol% of naphthenic acid in separate experiments. This shows that nonylphenol is a more effective surfactant than the acid, since a smaller quantity achieves a comparable dip in the optimal salinity. Our Method of Moments simulations reproduced these trends in the optimal salinity for all three crude components. We now have a coarse-grained molecular model that can handle simple oils doped with a single surface-active crude component. The next step is to (1) broaden the library of crude components, and (2) extend the framework to mixtures of molecules to predict shifts in the optimal salinity.

A molecular simulation technique like DPD relies on forcefields that characterise the interaction of different molecules with each other. It is therefore important to parametrise the interactions of the surface-active species with water and oil to mimic optimal salinities from phase behaviour experiments. The novelty of our method lies in developing these parameters starting from atomistic structures of the interacting species and predicting trends in a macroscopically observable quantity – the optimal salinity – that is crucial in implementing EOR in diverse fields.

434124 Rheological Study of Asphalt Using Molecular Simulation and Signal Processing

Abstract:

Asphalt is an amorphous material whose mechanical performance relies on viscoelastic responses to applied strain or stress. High quality pavement can enhance the commute experience and also can reduce the risk of accidents as well as preventing air pollution. Furthermore, it also can reduce the need for expensive reconstruction and maintenance of roads.

Having a broad understanding of asphalt chemistry is the key to understand and improve mechanical properties of this material. Chemical composition and its effect on the viscoelastic properties of model asphalts have been investigated by computing complex modulus from molecular dynamics simulation results for two different model asphalts (ZG08 and LG14) whose compositions each resemble the Strategic Highway Research Program AAA-1 asphalt in different ways.

Complex moduli at different temperatures have been calculated using fast Fourier transform to study the effect of temperature on viscoelastic properties of asphalt models. Because of inherent noise, any comparison was almost impossible. To remove the noise, signal processing techniques have been exploited. Signal processing techniques enhanced the clarity of the results and removed the noise from our results. LG14 have shown a very good agreement with the low and high frequency scaling limits of the Maxwell model within the frequency ranges spanned by the molecular dynamics simulations, while the ZG08 model asphalt only follow the high frequency scaling limits of the Maxwell model.

423805 Molecular Simulation of Organic Electrolyte Solutions

Abstract:

Molecular Simulation of Organic Electrolyte Solutions

Maximilian Kohns, Steffen Reiser, Martin Horsch, Hans Hasse

Laboratory of Engineering Thermodynamics (LTD), University of Kaiserslautern, Germany

Electrolyte solutions play an important role in many industrial processes. In the present work, we apply molecular simulation based on classical force fields with explicit solvent models to determine thermodynamic and structural properties of organic electrolyte solutions. Furthermore, the density of organic electrolyte solutions is studied experimentally. The focus of this work is on methanolic and ethanolic alkali halide salt solutions. The ion and solvent force fields are taken from previous work of our group. The ions are modeled as one Lennard-Jones (LJ) site with a point charge of ±1e in its center of mass [1,2]. Their force fields were optimized for describing basic thermodynamic data of aqueous solutions. The molecular models of the organic solvents methanol and ethanol consist of two and three LJ sites, respectively, and three partial charges to model both polarity and hydrogen bonding [3,4,5]. These models were adjusted to the vapor-liquid equilibrium of the pure component. The simulations were performed with the molecular simulation program *ms*2 [6]. The long range electrostatic interactions were considered by Ewald summation.

In the present work, basic thermodynamic properties of organic electrolyte solutions with the solvents methanol and ethanol are predicted by molecular simulation. All combinations of alkali cations and halide anions that are soluble in methanol and ethanol, respectively, are considered. The Lorentz-Berthelot combining rule is used for modeling the unlike LJ interactions between the ions and the solvent molecules. No adjustments to data for electrolytes in organic solvents are made. The predictions of the reduced liquid solution density from molecular simulation are in excellent agreement with experimental data, which are measured in the present work. This holds for all studied temperatures.

Furthermore, structural properties like the radial distribution function of the solvent molecules around the ions and transport properties like the self-diffusion coefficient and the electric conductivity of the organic electrolyte solutions are investigated.

[1] S. Deublein, J. Vrabec, H. Hasse, J. Chem. Phys. 136 (2012) 084501.

[2] S. Reiser, S. Deublein, J. Vrabec, H. Hasse, J. Chem. Phys. 140 (2014) 044504.

[3] T. Schnabel, J. Vrabec, H. Hasse, Fluid Phase Equilib. 239 (2006) 125-126.

[4] T. Schnabel, A. Srivastava, J. Vrabec, H. Hasse, J. Phys. Chem. B 111 (2007) 9871-9878.

[5] S. Reiser, N. McCann, M. Horsch, H. Hasse, J. Supercrit. Fluids 68 (2012) 94-103.

[6] C. W. Glass, S. Reiser, G. Rutkai, S. Deublein, A. Köster, G. Guevara-Carrion, A. Wafai, M. Horsch, M. Bernreuther, T. Windmann, H. Hasse, J. Vrabec, Comput. Phys. Commun. 185 (2014) 3302-3306.

429406 Using Atomistic Simulation to Determine the Optimal Aggregation Number of a Micelle, and Its Equilibrium with Polymer Surface in Aqueous Solution

Abstract:

In waterborne coatings, surfactants in the aqueous phase can either adsorb to polymer particle surfaces, or form micelles above the critical micelle concentration. The balance of surfactants between micelles and polymer surfaces is crucial for the colloidal stability of polymer particles. Here, we describe this balance by calculating the potential of mean force (PMF) to pull a sodium dodecyl sulfate (SDS) surfactant from a polymer surface and from a micelle respectively, using atomistic molecular dynamics simulations. For the PMF of the polymer surface, we investigated the role of salts in solution, other SDS molecules, and bound surface charges on the PMF, all of which influence the electrostatic forces between SDS and the surface. We calculated the PMF for SDS micelles with an aggregation number ranging from 10 to 100, from which the optimal aggregation number can be determined. This provides a practical method to calculate the aggregation number without simulating the exchange of surfactants between multiple micelles, which occurs at the time scale of seconds, and far surpasses the capability of atomistic simulations. Our predicted aggregation number of SDS micelles is in a very good agreement with the experimental data. By comparing the PMF of removing a SDS molecule from a polymer surface and from a micelle, we can predict the equilibrium of SDS molecules in the presence of both micelle and polymer surface without actually simulating both, which reduces computational cost and provides guidance for formulation of coatings.

423755 Pseudo Phase Diagrams of Supported Lipid Bilayers Formation: A Density Functional Theory Study

Abstract:

Pseudo phase diagrams of supported lipid bilayers formation: a density functional theory study

Xian Kong[1], Diannan Lu[1], Jianzhong Wu[2], Zheng Liu[1]

[1] Department of Chemical Engineering, Tsinghua University, Beijing 100084, China

[2]Department of Chemical and Environmental Engineering and Department of Mathematics, University of California, Riverside, CA 92521, USA

Solid supported lipid bilayers (SLBs) maintain the structural and dynamical properties of natural lipid bilayer membranes promising for a wide variety of applications including biosensors and biomimetic membranes[1-3]. A molecular theory for predicting SLB formation from different lipid solutions and substrates is instrumental for engineering design, development and applications[4-6]. Whereas the structure of freestanding lipid bilayers has been the subject of longstanding research, relatively few theoretical studies have been devoted to examine how the supporting solid affect the bilayer structure and phase behavior over various conditions. In this work, we investigate the SLB's formation possibility using a coarse-grained model and the classical density functional theory (DFT) that accounts for the molecular topology, electrostatic correlations and molecular size effects[7,8]. The phase diagrams of supported lipid bilayers in different supporting substrates depend on lipid properties including lipid tail length, lipid head size and charge of the polar head, as well as substrate properties such as surface charge density and hydrophobicity.

References:

1 Sackmann, E. Supported membranes: Scientific and practical applications. *Science* **271**, 43-48, doi:DOI 10.1126/science.271.5245.43 (1996).

2 Kaufman, Y., Berman, A. & Freger, V. Supported Lipid Bilayer Membranes for Water Purification by Reverse Osmosis. *Langmuir* **26**, 7388-7395, doi:Doi 10.1021/La904411b (2010).

3 Castellana, E. T. & Cremer, P. S. Solid supported lipid bilayers: From biophysical studies to sensor design. *Surface Science Reports* **61**, 429-444, doi:DOI 10.1016/j.surfrep.2006.06.001 (2006).

4 Richter, R. P. & Brisson, A. R. Following the formation of supported lipid bilayers on mica: A study combining AFM, QCM-D, and ellipsometry. *Biophysical Journal* **88**, 3422-3433, doi:DOI 10.1529/biophysj.104.053728 (2005).

5 Mulligan, K., Jakubek, Z. J. & Johnston, L. J. Supported Lipid Bilayers on Biocompatible Polysaccharide Multilayers. *Langmuir* **27**, 14352-14359, doi:Doi 10.1021/La203207p (2011).

6 Garcia-Manyes, S., Oncins, G. & Sanz, F. Effect of pH and ionic strength on phospholipid nanomechanics and on deposition process onto hydrophilic surfaces measured by AFM. *Electrochimica Acta* **51**, 5029-5036, doi:DOI 10.1016/j.electacta.2006.03.062 (2006).

7 Wu, J. Z. & Li, Z. D. Density-functional theory for complex fluids. *Annu Rev Phys Chem* **58**, 85-112, doi:DOI 10.1146/annurev.physchem.58.032806.104650 (2007).

8 Li, Z. & Wu, J. Density functional theory for polyelectrolytes near oppositely charged surfaces. *Phys Rev Lett* **96**, 048302 (2006).

424914 Prediction of Vapor and Sublimation Pressure for Pure Compounds from the Modified PR+Cosmosac Equation of State

Abstract:

Vapor pressure for pure liquids and solids is an important physical property for many applications in chemical engineering process. Although they can be obtained by experiment whenever possible, there are cases when direct measurement is not allowed, for example, for chemically unstable species in the pure state and for newly synthesized species where often very limited amount of quantity is available. It is therefore desirable to have a predictive tool prior to measurement. In this study, the PR+COSMOSAC EOS is used for such a purpose. In particular, two modifications are introduced to increase the accuracy of the prediction near and below the triple point. The overall average logarithmic deviation for vapor pressure (ALD-*P*) for 1124 pure liquids from the modified version of PR+COSMOSAC EOS is 0.229 (138%), which is significantly lower than that from the original version (553%). In particular, the ALD-P for the prediction of vapor pressure near triple point reduces from 0.391 (1062%) to 0.321 (233%). When the melting properties are available, the sublimation pressure can also be estimated. The ALD-P for the sublimation pressure for 1140 substances shows great improvement, from 1.13 (1249%) to 0.71 (413%). These properties predicted from PR+COSMOSAC EOS can further be used for the solubility of drugs in various kinds of solvent. The modified PR+COSMOSAC EOS can provide useful information while the experiment data is not available

414754 Metal-Organic Frameworks for Liquid-Phase Applications: A Computational Perspective

Abstract:

As a unique class of hybrid nanoporous materials, metal-organic frameworks (MOFs) have received tremendous interest over the last decade. The variation of metal oxides and the judicious choice of controllable organic linkers allow the pore size, volume and functionality to be tailored in a rational manner for designable architectures. MOFs thus provide a wealth of opportunities for engineering new functional materials and have been considered as versatile candidates for many important potential applications. This presentation will give an overview of recent computational studies, by our research group, using MOFs for liquid-phase applications such as water desalination, biofuel purification, chiral separation and biomass recovery.

425998 Modern Catalytic Technologies for Converting Biomass to Renewable Fuels and Chemicals

Abstract:

In this talk, an overview of modern catalytic-based technologies will be presented that may overcome challenges of traditional methods in converting biomass to renewable fuels and chemicals. Diels-Alders and dehydration chemistry will be outlined toward the production of renewable aromatics, such as para-xylene.

409094 Molecular-Level Kinetic Modeling in Thermochemical Conversions: Software Tools and Their Applications

Abstract:

The considerable interest in molecule-based models of the structure and reactions of petroleum, biomass and other unconventional energy resources is motivated by the need to predict reaction properties. This is because the molecular composition is an optimal starting point for the prediction of mixture properties. The potential advantages of molecule-based modeling are thus clear. Less readily apparent, however, is that the development and operation of molecular models comes with a large requirement for model construction and solution time as well as analytical and reactivity information.

The challenge of building molecule-based models is due to the staggering complexity of the complex reaction mixtures. There will often be thousands of potential molecular and intermediate (e.g., ions or radicals) species. The sheer size of the thus-implied modeling problem creates a conflict between the need for molecular detail and the formulation and solution of the model. Clearly, the use of the computer to not only solve but also formulate the model would be helpful in that it would allow the modeler to focus on the basic chemistry, physics and approximations of the model.

Our recent work has led to the development of the Kinetic Modeler's Toolbox (KMT), which organizes software tools for an automated construction, solution and optimization of detailed kinetic models. KMT's Composition Modeling Editor (CME) transforms bulk analytical chemistry into a molecular description of the feedstock. This casts the modeling problem in molecular terms. Reactivity information is then organized in terms of quantitative linear free energy relationships. KMT's INGen tool allows for the model equations to be built and coded on the computer. KMT's Kinetic Model Editor (KME) allows for the solution of this chemical reaction network, in the context of the chemical reactor, which provides a prediction of the molecular composition that can then be organized into any desired commercially relevant outputs.

This approach is illustrated through the development of molecule-based models for refinery units and biomass pyrolysis and hydrotreating.

434505 Simultaneous Determination of Liquid Dynamic Viscosity and Density of MEA and DEA Aqueous Solutions at Temperatures Between (313-353) K and Pressures up to 30 Mpa

Abstract:

The removal of acid gas impurities such as CO_2 and H_2S from natural, refinery, and synthesis gas streams is an important operation in gas processing. Aqueous solutions of alkanolamine as chemical absorbents have been used in the removal of acid gases for quite a long time, and they are still widely used nowadays. Industrially important alkanolamines are monoethanolamine (MEA), diethanolamine (DEA). The physical properties of solvents such as density and dynamic viscosity are those data sets that are required and important for designing or simulation of an absorption process for CO_2capturing or separation. Data of these thermophysical properties at atmospheric pressure are common, however, there is a lack of data at high pressures. For this reason, in the thermodynamic laboratory from Instituto Politécnico Nacional in Mexico City, a systematic study of behavior of dynamic viscosity and density of pure alkanolamines and its mixtures at temperature up to 353 K and pressures up to 30 MPa is being carried out. In this work, we reported simultaneous determinations of dynamic viscosity and density of MEA and DEA aqueous solutions at temperatures between (313-353) K and up to 30 MPa of pressure. The combined expanded uncertainties were evaluated to be 0.7 % for dynamic viscosity and 0.06 % for density.

Keywords: Dynamic viscosity at high pressures, MEA, DEA, density at high pressures.

* To whom correspondence should be addressed Luis A. Galicia-Luna, **E-mail:** lgalicial@ipn.mx

425122 Ionic Liquids - Phase Behavior to Applications

Abstract:

The history of ionic liquids (ILs) effectively started in 1914, when the physical properties of ethylammonium nitrate ($[CH_3CH_2NH_3^+][NO_3^-]$, m.p. 13-14 °C) were first reported. ILs are generally defined as salts composed of discrete cations and anions with melting points below 100 °C, and many are liquid at ambient temperature. IL research has grown rapidly over the past decade due to the realization that these materials have many unique properties such as negligible vapor pressure and outstanding solvation potential. Ionic liquids have been further emphasized by the fact that their physical and chemical properties can be finely tuned by varying both the cation and anion.

Our research has focused on accurately measuring vapor-liquid equilibria (VLE) and vapor-liquid-liquid equilibria (VLLE) and using thermodynamic models to understand the phase behavior of binary gas mixtures in ILs. This presentation will focus on the importance of characterizing the global phase behavior of gases in ionic liquids and how this can provide insight into new applications. Solubility measurements of several gases in ILs will be discussed and important experimental details regarding VLE measurements using a gravimetric microbalance and VLLE measurements using a mass-volume technique will be highlighted. VLE data have been successfully correlated with a modified Redlich-Kwong equation of state (EOS), and in certain cases (*e.g.* hydrofluorocarbons) the EOS predicts partial immiscibilities (LLE) with lower critical solution temperatures (LCSTs) in the fluorocarbon-rich side solutions. We have also found that gases such as CO_2 can exhibit different solubility behaviors in ILs (*i.e.* physical and chemical absorption) and that these behaviors can be analyzed with the EOS using a simple association model and excess thermodynamic functions.

Knowledge of gas and IL phase behavior has led to several practical applications including separation of azeotropic gas mixtures, removal of CO_2 from flue gas and new fluid pairs for absorption cooling. These along with other interesting examples from the literature will be discussed.

419950 Developing Quantum Mechanically-Based Force Field Parameters from Gas Hydrates to Biology

Abstract:

Prof. Stanley Sandler was one of the pioneers in the chemical engineering field to utilize quantum mechanical (QM) methods for applications to phase equilibria via thermodynamic models and developing force field parameters for use in molecular simulations. I was fortunate to work under Prof. Sandler during this time period as a Ph.D. student in developing guest-host interaction potentials in gas hydrates. In this talk, I will summarize how QM was used to develop accurate guest-water potentials for use in predicting gas hydrate phase equilibria and guest adsorption in its water lattice. This fundamental approach to force field development was extended to a more complex class of biomolecules, i.e., lipids. QM methods are currently used in predicting torsional energy profiles that are then fit to potential functions in complex biomolecular force fields. Moreover, QM has been used to guide fits for the partial charge distribution and other non-bonded interactions in lipid force fields. The high accuracy of the CHARMM36 pairwise additive lipid force field will be presented and a prospective on future lipid force field development needs will also be be included.

423779 Fundamentals and Applications of Antibody Solution Thermodynamics

Abstract:

There is intense interest in using monoclonal antibodies as therapeutics because of their unique properties and activities. Nevertheless, there are many challenges in developing therapeutic antibodies related to their variable and difficult-to-predict propensities to self-associate, which can lead to a range of undesirable solution properties such as low solubility and abnormally high viscosity. I will discuss our recent progress in using self-interaction and phase behavior measurements of antibodies to understand and optimize their sequence-dependent thermodynamic properties.

430952 Thermodynamic Properties of Monoclonal Antibody Solutions

Abstract:

Monoclonal antibodies (mAbs) are a class of proteins that represent the largest single group of protein drugs produced by the pharmaceutical and biotechnology industries; annual sales of mAbs now exceed $100 billion. All mAbs share extensive common structural features, but very small local changes in structure, including substitution of a small number (1-3) of amino acids, can lead to dramatic changes in thermodynamic and transport properties. This presentation will cover examples of such changes via a variety of different measurements, with a special focus on the effects of small structural changes. These include effects on mAb self-interactions, oligomerization, phase behavior, and viscosity. Potential molecular bases for this sensitivity will be examined.

432803 Multiscale Modeling for DNA Simulations

Abstract:

Numerous experimental and theoretical studies have offered clues regarding the molecular recognition capabilities of DNA, but our understanding of many aspects of the self-assembly process is still far from complete. The goal of our DNA model is to help improve the fundamental understanding of the physical principles that underlie the DNA self-assembly process and to provide a computational tool based on molecular-level simulations that will enhance our ability to study the formation of DNA based nanostructures. We used a multiscale modeling approach to develop an implicit-solvent intermediate-resolution model for DNA molecules designed for use with discontinuous molecular dynamics simulations and applied to study DNA structural properties and the hybridization process in solution. The geometric and energetic parameters for this model were obtained by collecting data from atomistic simulations of single stranded DNA molecules with explicit solvent and counterions. In this model, the sugar, phosphate and base in each nucleotide are each represented by a single coarse-grained site. Using this new model we are able to simulate the spontaneous hybridization of 2 DNA strands composed of 24 base pairs in minutes. Simulation results show that the model accurately reproduces the hybridization of complementary DNA strands in mixtures containing both complementary and non-complementary DNA strands. Current work is focused on using the model to predict the formation of Holliday Junctions.

Qatar Carbonates and Carbon Storage Research Centre

Diffusion Coefficients of Carbon Dioxide in Liquid Hydrocarbons at High Pressures: Experiment and Modeling

Shane Cadogan, Geoffrey C. Maitland, Bhavik Mistry, J. P. Martin Trusler and Timothy Wong

Department of Chemical Engineering

Qatar Carbonates and Carbon Storage Research Centre

Imperial College London, U.K.

Stanley Sandler 75th Birthday Session

AIChE Annual Meeting, Salt Lake City, 9 November 2015

Imperial College London | Qatar Petroleum | QATAR SCIENCE & TECHNOLOGY PARK

Qatar Carbonates and Carbon Storage Research Centre

Outline and Motivation

Outline

- Experimental method
- Results for CO_2 diffusion in six normal alkanes and one branched alkane
- Rough-hard-sphere theory for mutual diffusion
- Towards a universal correlation

Motivation

- Diffusion coefficients are important in estimating interfacial mass transfer in many industrial processes
- Significant in the context of carbon storage and CO_2-EOR in oil reservoirs
- Little date available at high temperatures and pressures

Imperial College London | Qatar Petroleum | QATAR SCIENCE & TECHNOLOGY PARK

2

Qatar Carbonates and Carbon Storage Research Centre

Experimental Method: 1. Taylor Dispersion Apparatus

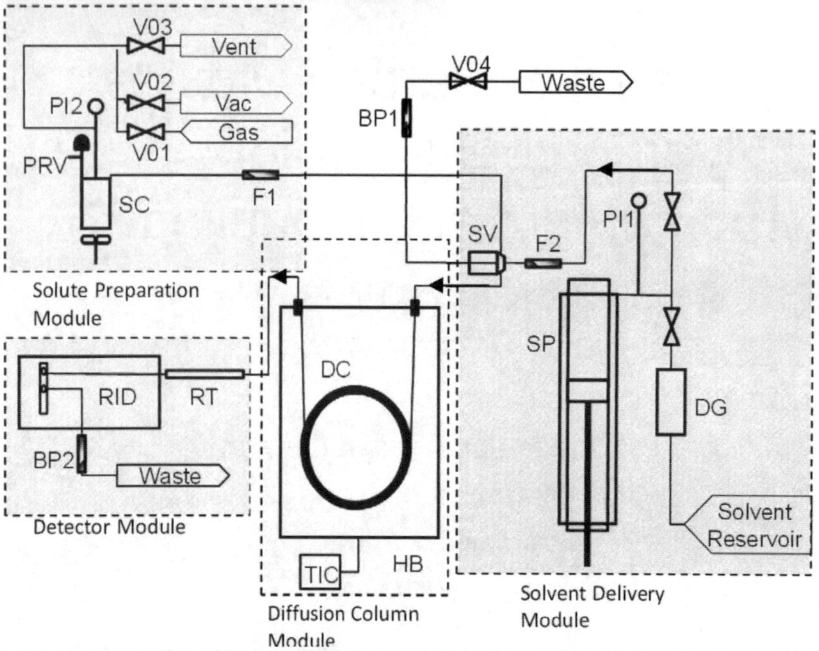

$298 < T/\mathrm{K} < 423$

$p/\mathrm{MPa} \leq 69$

$[\mathrm{Re}] < 30$

$[\mathrm{De}]^2[\mathrm{Sc}] < 20$

Qatar Carbonates and Carbon Storage Research Centre

Experimental Method: 2. Taylor Dispersion Principle

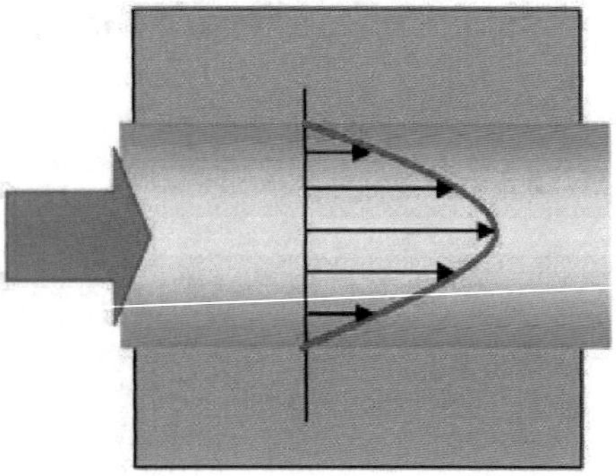

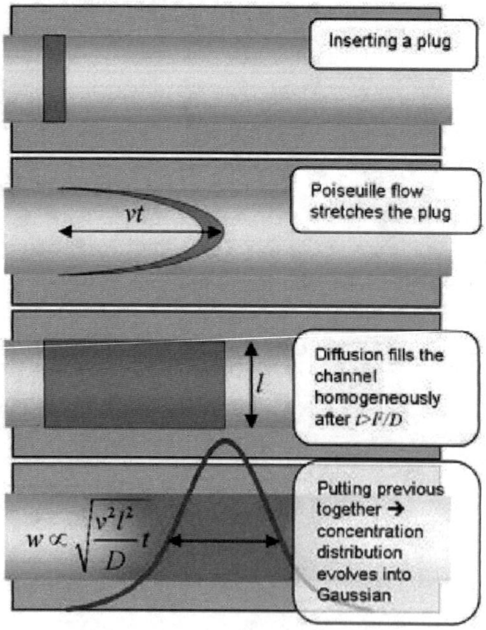

Qatar Carbonates and Carbon Storage Research Centre

Experimental Method: 3. Working Equation

- Refractive index detection:
  $s(t) = a + bt + \alpha c(t)$
- *a*, *b* baseline coefficients
- α detector sensitivity
- Extended Aris model:

$$c(t) = \left\{ \frac{n}{\pi R^2 \sqrt{4\pi K t}} \right\} \exp\left[-\frac{(L - vt)^2}{4Kt} \right]$$

$$K = D + \frac{R^2 v^2}{48D}$$

$$L = L_c + \left(\frac{D}{D_0} \right) \sum_{i=1}^{3} \left(\frac{R_i}{R_c} \right) L_i$$

- Typical standard relative uncertainty of 2 %

Example signal: CO_2 in octane at T = 298 K and p = 1 MPa

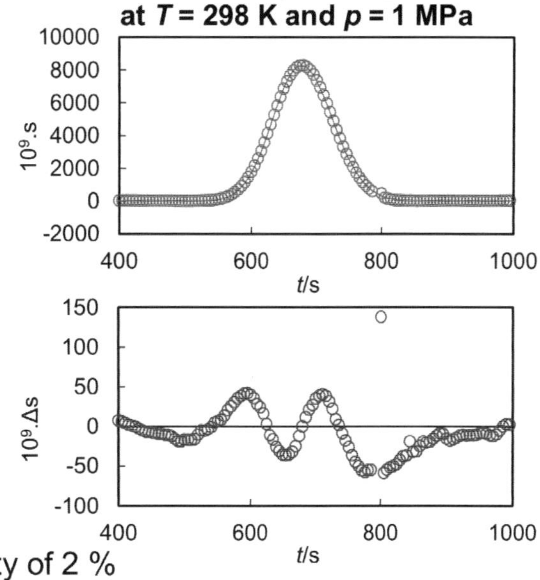

Imperial College London

قطر للبترول Qatar Petroleum

QATAR SCIENCE & TECHNOLOGY PARK

5

Qatar Carbonates and Carbon Storage Research Centre

Experimental Results: (298 to 423)K and (1 to 69) MPa

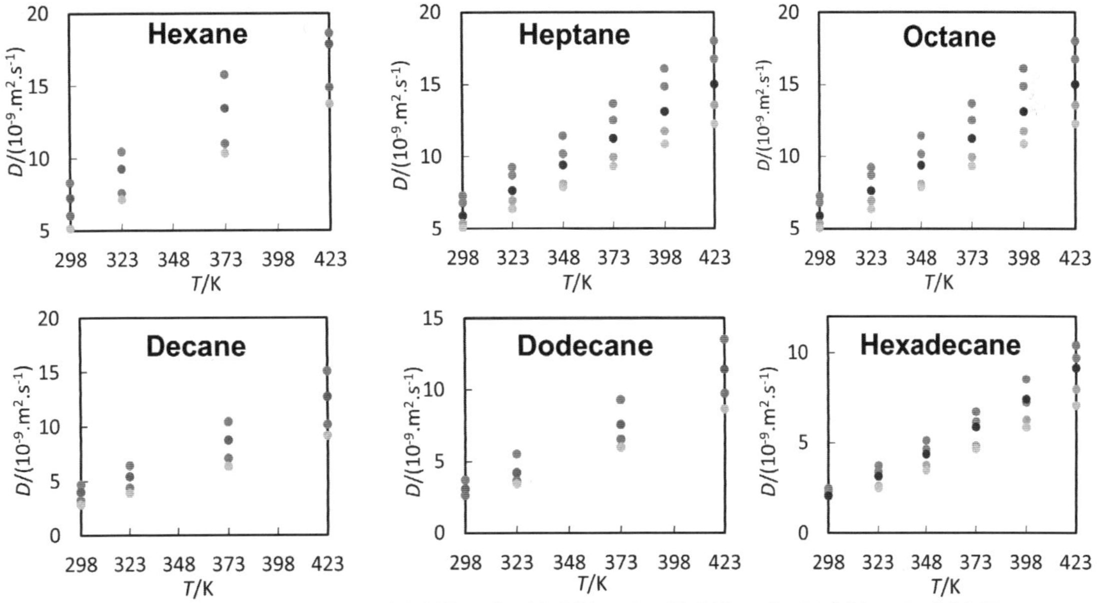

●, 1 MPa; ●, 10 MPa; ●, 23 MPa; ●, 30 MPa; ●, 46 MPa; ●, 50 MPa; ●, 69 MPa

Imperial College London

قطر للبترول Qatar Petroleum

QATAR SCIENCE & TECHNOLOGY PARK

6

Qatar Carbonates and Carbon Storage Research Centre

Experimental Results: (298 to 423)K and (1 to 69) MPa

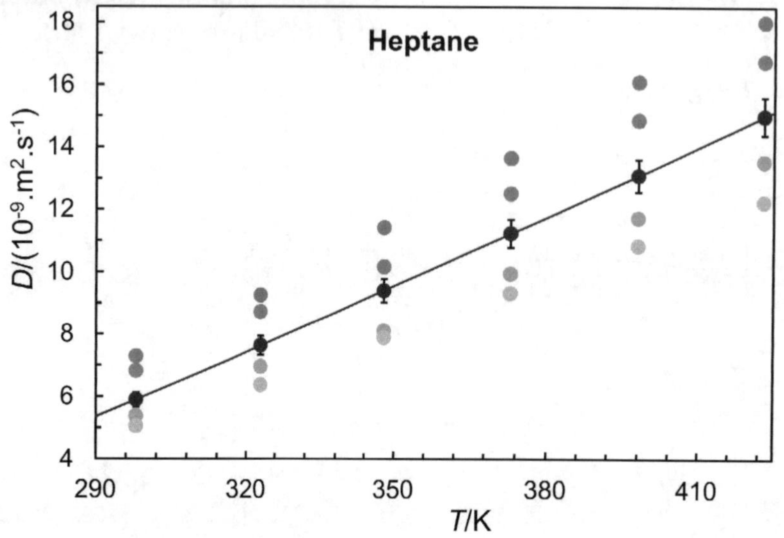

●, 1 MPa; ●, 10 MPa; ●, 30 MPa; ●, 50 MPa; ●, 69 MPa

Qatar Carbonates and Carbon Storage Research Centre

Rough-Hard-Sphere Model for Binary Diffusion at $x_1 \to 0$

- Dilute gas of smooth hard spheres: $$D_{12}^0 = \frac{3}{8n\sigma_{12}^2}\sqrt{\frac{kT}{\pi\mu_{12}}}$$
- Dense gas of rough hard spheres: $$D_{12} = \frac{A_{12}D_{12}^0}{g(\sigma_{12})}$$
- Liquid of rough hard spheres: $$D_{12} = \frac{A_{12}D_{12}^0}{g(\sigma_{12})}\cdot C(\sigma_1,\sigma_2,m_1,m_2,V)$$
- From MD studies: $$\frac{C(\sigma_1,\sigma_2,m_1,m_2,V)}{g(\sigma_{12})} = \frac{c(V-V_D)}{V}$$
- Final expression: $$\boxed{D_{12} = \beta\sqrt{T}(V-V_D)}$$

Qatar Carbonates and Carbon Storage Research Centre

Comparison with Experiment - normal alkanes

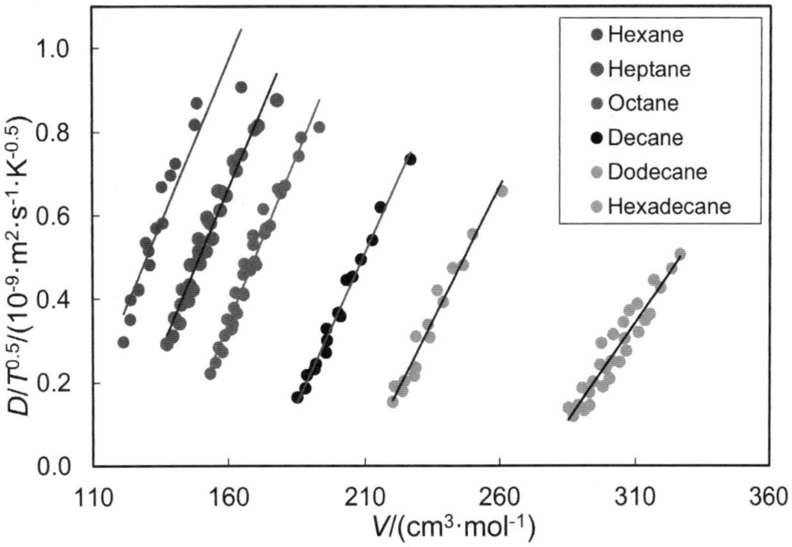

Imperial College London | قطر للبترول Qatar Petroleum | QATAR SCIENCE & TECHNOLOGY PARK

9

Qatar Carbonates and Carbon Storage Research Centre

Universal Correlation

- From RHS theory, each solute-solvent system is characterized by two parameters: β and V_D
- V_D is the molar volume of the solvent at which diffusion ceases
- Hypothesize that V_D is proportional to V_0, the molar core volume of the pure solvent, and that βV_0 is constant for a given solute in a homologous series of solvents
- In that case, we have a universal expression for the diffusion coefficient of a given solute gas in a homologous series of solvents:

$$\frac{D_{12}}{\sqrt{T}} = a\left(\frac{V}{V_0} - b\right)$$

- For normal alkanes, viscosity data provides an expression for V_0 in the form:

$$V_0 = f(C_n, T)$$

C_n = carbon number

Imperial College London | قطر للبترول Qatar Petroleum | QATAR SCIENCE & TECHNOLOGY PARK

10

Qatar Carbonates and Carbon Storage Research Centre

Correlation for CO_2 Diffusion in normal Alkanes

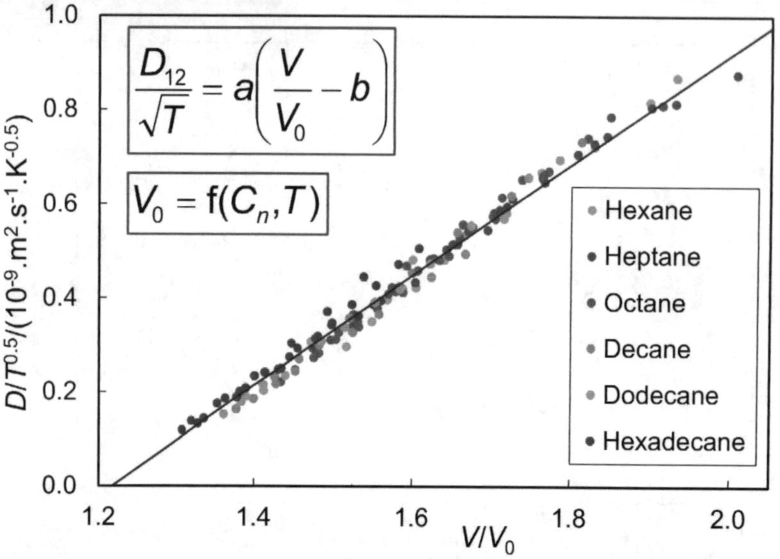

a = 1.173	b = 1.217	Δ_{AAD} = 5 %	1.3 ≤ V/V_0 ≤ 2.0

Imperial College London | قطر للبترول Qatar Petroleum | QATAR SCIENCE & TECHNOLOGY PARK | 11

Qatar Carbonates and Carbon Storage Research Centre

Results for Squalane
(2,6,10,15,19,23-hexamethyltetracosane)

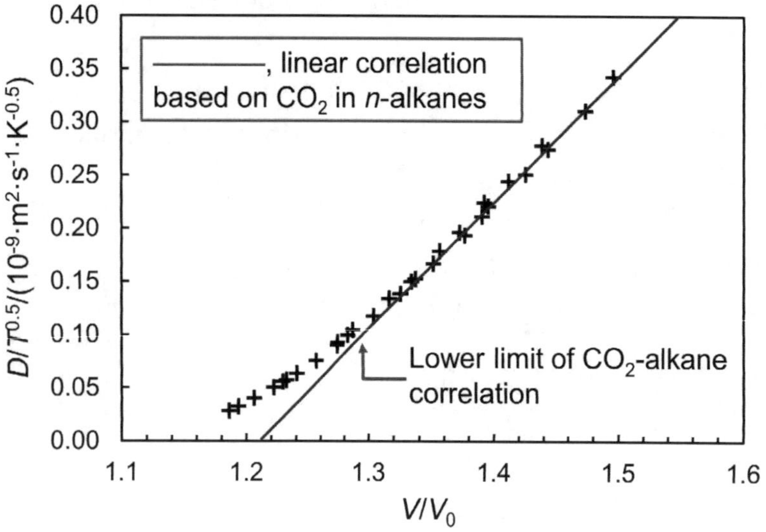

Results suggests that the assumption of a universal correlation may be good but that the constraint of a *linear* relationship should be relaxed.

Imperial College London | قطر للبترول Qatar Petroleum | QATAR SCIENCE & TECHNOLOGY PARK | 12

Qatar Carbonates and Carbon Storage Research Centre

Summary and Acknowledgement

Summary

In this work we have:

- Obtained new experimental data for CO_2 diffusion in normal alkanes from C_6 to C_{16} and in squalane ($C_{30}H_{62}$)
- Developed a universal correlation for the *n*-alkane systems in terms of temperature, solvent molar volume and carbon number
- Squalane data suggests that the correlation should become non-linear at high densities

Acknowledgement

Stanley Sandler: inspiring, leading, educating. Thank you!

This work was carried out as part of the activities of the Qatar Carbonates and Carbon Storage Research Centre (QCCSRC). We gratefully acknowledge the funding of QCCSRC provided jointly by Qatar Petroleum, Shell, and the Qatar Science and Technology Park, and their permission to publish this research.

Imperial College London   قطر للبترول Qatar Petroleum   QATAR SCIENCE & TECHNOLOGY PARK   13

Qatar Carbonates and Carbon Storage Research Centre

Comparison: Experiment vs Universal Correlation

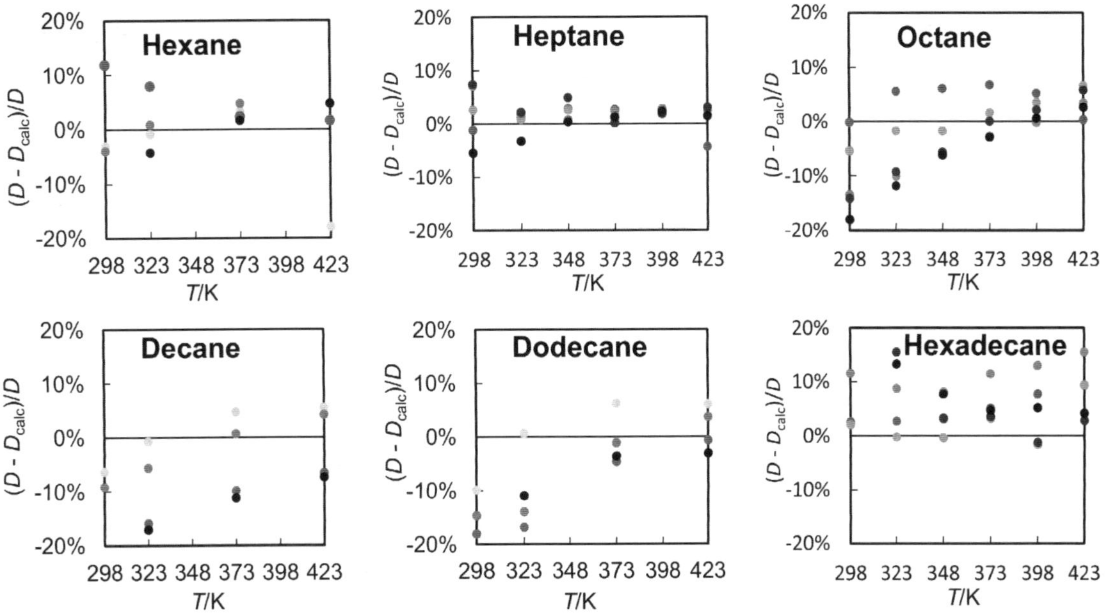

Imperial College London   قطر للبترول Qatar Petroleum   QATAR SCIENCE & TECHNOLOGY PARK   14

437483 Designing Software for Chemical Engineers – a Product Developer's Perspective

Abstract:

Process simulators are tools widely used for the design of chemical processes and chemical products. These products are ubiquitous in process engineering work and a normal part of the engineer's work flow. These tools and associated tools like the software plug-ins designed to provide highly specialized services such as mass transfer separation modules, flare systems and advanced chemical reactors. Although the tools are widely used little is said about the process used to develop these tools, the challenges the software product developer faces and how these challenges are met. In this presentation we take an introspective look at the development of software products commonly used in process design and process simulation.

425315 Designing Chimeric Peptides Using Swarm Intelligence Optimization for the Protein Folding Problem

Abstract:

Chimeric peptides offer biological engineering methodologies adapted in materials science for wide range of applications including assembly and synthesis. Tailored chimeric functionality depends not only the sequence but also the molecular structure of the resulting molecules. Designing new chimeric peptides therefore relies on developing a better understanding of the protein folding. Here, a new *ab initio* method for finding peptide backbone angles was developed with a swarm intelligence algorithm to find peptide folds faster than using a molecular dynamics approach. Currently, many approaches to the protein folding problem have been investigated. The first approach is to experimentally determine the structures. This is the most accurate way of determining the structure, but the process is slow and requires commitments of large resources. Determining a structure experimentally without knowing that the peptide design is already successful is prohibitive. The second approach is to use molecular dynamics simulations. The main limitation of these simulations for peptide design is the low number of peptides that can be analyzed. Having larger peptide design candidate pools should help to improve chimeric peptide design. Using existing data avoids the cost of determining new structures and statistical analysis avoids the computational cost of simulating the protein folding process in the second approach. However in swarm intelligence algorithm, statistical analysis of determined folding structures may avoid the challenges of the first two approaches. In previous work[1], an antimicrobial peptide that self-assembles onto titanium implants has been developed. Here, we applied two optimization strategies to these peptide sets for finding chimeric peptide folds more effectively. We next compared their folding patterns to engineer sequence-structure function relations. The first optimization strategy uses a swarm intelligence algorithm inspired by bee hive coomunication to coordinate the selection of highly likely folds. The second approach is to match the peptide fragments with historical data of protein folds[2]. PyRosetta software[3] was used to score the likelihood of structures and to find the bond angles for the amino acid side chains. The likelihood of the structures for each approach was compared, and a time analysis was performed. Our approach offers to expand the pool of new chimeric peptide design. *The NIH-NIAMS (7R21AR062249-03) is acknowledged.*

[1] Yucesoy D, Hnilova M, Boone K, Arnold P, Snead M, Tamerler C. Chimeric Peptides as Implant Functionalization Agents for Titanium Alloy Implants with Antimicrobial Properties. JOM 2015;67:754-66.

[2] Kim DE, Chivian D, Baker D. Protein structure prediction and analysis using the Robetta server. Nucleic Acids Research 2004;32:W526-W31.

[3] Chaudhury S, Lyskov S, Gray JJ. PyRosetta: a script-based interface for implementing molecular modeling algorithms using Rosetta. Bioinformatics 2010;26:689-91.

15AIChE
Annual Meeting, Salt Lake City, UT

DTU

A New Optimization Model for Computer-Aided Molecular Design Problems

Lei Zhang, **Stefano Cignitti**, Rafiqul Gani*

PSE for SPEED Project, Department of Chemical and Biochemical Engineering, Technical University of Denmark

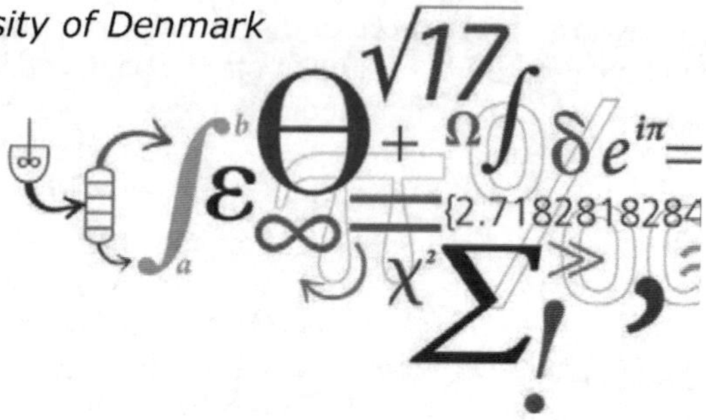

DTU Chemical Engineering
Department of Chemical and Biochemical Engineering

15AIChE
Annual Meeting, Salt Lake City, UT

DTU

Outline

- Introduction
- Computer-aided Framework for Product Design Problems
- New Optimization Model for CAMD
- Case Studies
  - Surfactant Design
  - Refrigerant Design
  - Die Attach Adhesive Design
- Conclusion

15AIChE
Annual Meeting, Salt Lake City, UT

DTU

Chemical Product Design

- The modern society needs many products for its survival.
- Which are the important products, which resources should be used to produce them and what is the role of the chemicals based manufacturing process?
- With the help of computer-aided chemical product design, the optimal, more sustainable product-process will be designed.

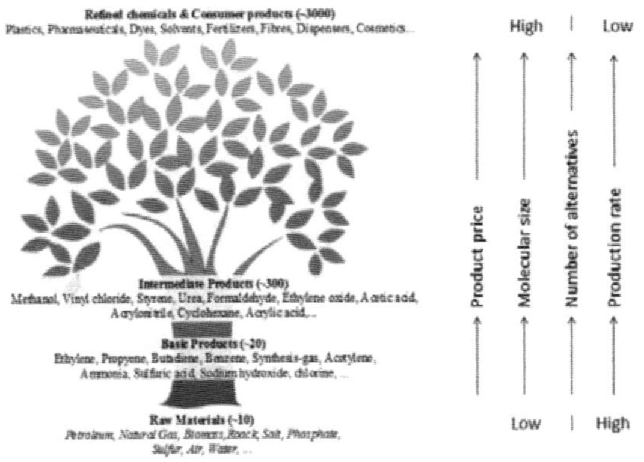

15AIChE
Annual Meeting, Salt Lake City, UT

DTU

Computer-Aided Chemical Product Design

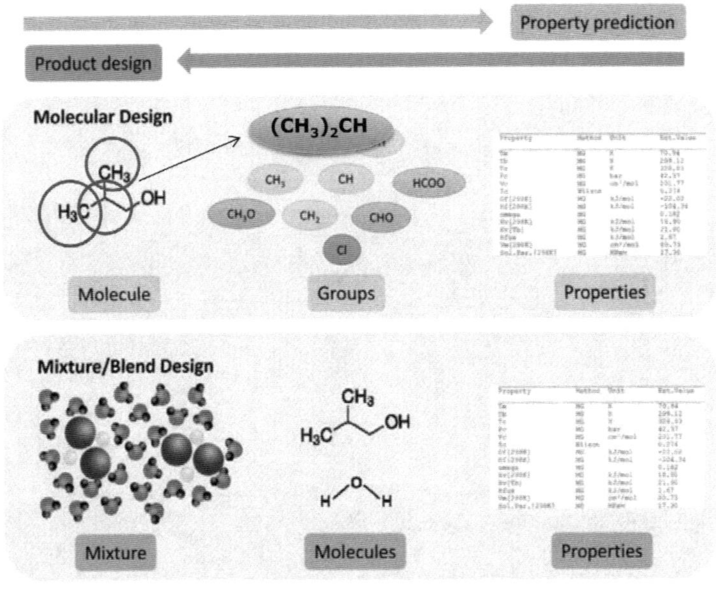

Computer-Aided Framework for Product Design Problems

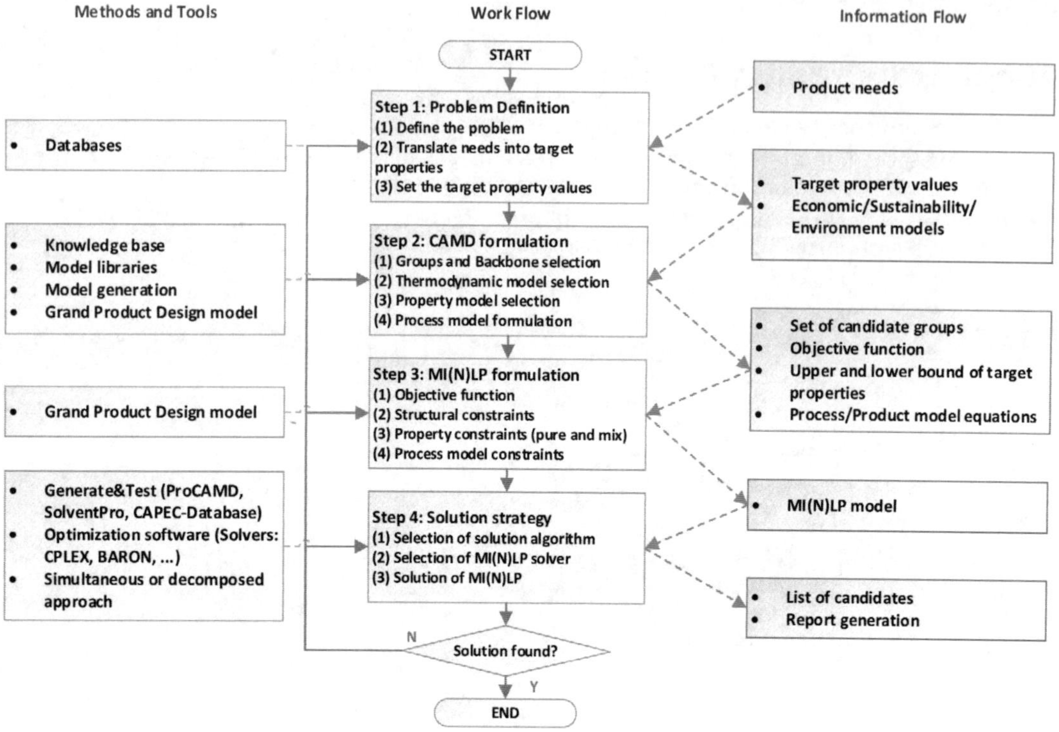

New Optimization Model for CAMD: MINLP Formulation

$$\min / \max f_{obj}(\mathbf{X}, \mathbf{N})$$

Structural constraints: $g_1(\mathbf{N}, \mathbf{Y}) \leq 0$

Pure component property: $g_2(\mathbf{N}, \mathbf{Y}) \leq 0$

Mixture property: $g_3(\mathbf{X}, \mathbf{N}) \leq 0$

Process model and others: $g_4(\mathbf{X}, \mathbf{N}) = 0$

$$\mathbf{X} \in \mathbb{R}^n; \mathbf{N} \in \mathbb{Z}_+^m; \mathbf{Y} \in \{0,1\}^q$$

$\mathbf{N}$: group number;

$\mathbf{Y}$: adjacency matrix;

$\mathbf{X}$: continuous variables related to the mixture and process.

- Adjacency matrix is introduced to predict the detailed structure of the molecule.
- New constraints are involved to find the number of 2nd order groups from the adjacency matrix using binary variables.

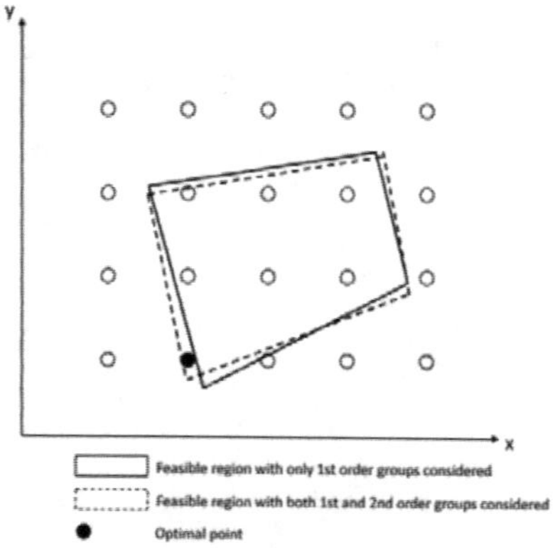

Figure: Higher order groups need to be considered simultaneously to obtain "real" optimal product

15AIChE
Annual Meeting, Salt Lake City, UT

DTU

Generic mathematical programming formulation

1. Structural constraints:

1.1 Octet rule

$$\sum_{i \in G_1} (2 - \nu_i)\, n_i^{(1)} = 2q$$

$$\sum_{i_1 \neq i_2; i_1 \in G_1} n_{i_1}^{(1)} \geq n_{i_2}^{(1)} (\nu_{i_2} - 2) + 2 \quad \forall i_2 \in G_1$$

1.2 Constraints ensure only one molecule is formed (Churi and Achenie (1996)).

1.3 Group number constraints:

$$n_i^L \leq n_i^{(1)} \leq n_i^U \quad \forall i \in G_1$$

$$n^{min} \leq \sum_{i \in G_1} n_i^{(1)} \leq n^{max}$$

1.4 Adjacency matrix constraints:

$$y_{i_1,id_1,i_1,id_1} = 0 \quad \forall i_1 \in G_1, id_1 \in ID$$

$$y_{i_1,id_1,i_2,id_2} = y_{i_2,id_2,i_1,id_1} \quad \forall i \in G_1; id \in ID$$

$$\sum_{id_1 \in ID} z_{i_1,id_1} = n_{i_1}^{(1)} \quad \forall i_1 \in G_1$$

$$\sum_{i_2 \in G_1} \sum_{id_2 \in ID} y_{i_1,id_1,i_2,id_2} = \nu_{i_1} z_{i_1,id_1}$$

$$\forall i_1 \in G_1, id_1 \in ID$$

1.5 Second-order group constraints:

$$N_B^J - M\left(1 - b_{id_1,id_2,\cdots,id_{N_B}^J}\right) \leq \sum_{\text{If these two grps connect in 2nd grp. J}} y_{i_1,j_1,i_2,j_2} \leq N_B^J - 1 + M\left(b_{id_1,id_2,\cdots,id_{N_B}^J}\right) \quad \forall J \in G_2$$

$$n_j^{(2)} = \frac{1}{T} \sum_{id_1,id_2,\cdots,id_{N_B}} b_{id_1,id_2,\cdots,id_{N_B}^J} \quad \forall J \in G_2$$

2. Property constraints:

$$p_k^L \leq p_k \leq p_k^U \quad \forall k \in P$$

$$p_k = \sum_{i \in G_1} n_i^{(1)} p_k^{(1)} + \sum_{j \in G_2} n_j^{(2)} p_k^{(2)} \quad \forall k \in P$$

3. Process model, economic model and other constraints.

[1]Lei Zhang, Stefano Cignitti, Rafiqul Gani. *Computers & Chemical Engineering*, **78**, 79-84

Churi N, Achenie LE., Ind Eng Chem Res 1996;35(10):3788–94.

15AIChE
Annual Meeting, Salt Lake City, UT

DTU

Surfactant Design - Product Design (1/3)

- Step 1: Design of a UV sunscreen, in the emulsified form, with a high sun protection factor. The phenol ring is fixed as backbone structure in this case study.
- Step 2:
  - **Objective:** Toxicity should be minimized.
  - **Group selection:** CH_3 , CH_2 , CH , C , aCH , aC-OH , CH_2COO , CH_3O , CH_2O , aC-O , OCH_2CH_2OH from 220 groups (Marrero and Gani, (2001)).
  - **Backbone selection:** -C_6H_4OH

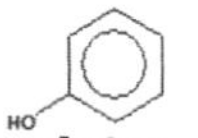

  - **Structural and property constraints:**

Need	Lower bound	Upper bound
N_G	10	15
N_F	0	8
$lc50$	3.16log(mol/L)	-
Sp	-	$25MPa^{0.5}$
clp	343.15K	-
Fp	343.15K	-
Vm	0.1cc/mol	0.3cc/mol

LC_{50}: Fathead Minnow 96-hr LC_{50}; *Sp*: Hildebrand solubility parameter; *clp*: cloud point; *Fp*: Flash point; *Vm*: Liquid molar volume at 298K.

15AIChE
Annual Meeting, Salt Lake City, UT

DTU

Surfactant Design (2/3)

- Step 3:

$$\max lc50 = lco + \sum_{i_1} n_{i_1}^{(1)} lc50_{i_1}^{(1)} + \sum_{i_2} n_{i_2}^{(2)} lc50_{i_2}^{(2)}$$

1. Structural constraints:

1.1 Octet rule and group number constraints:

$$\sum_{i} (2 - \nu_{i_1}) n_{i_1}^{(1)} = 0$$

$$\sum_{i_1, i_1 \neq i_1'} n_{i_1}^{(1)} \geq n_{i_1'}^{(1)} \left(\nu_{i_1'} - 2\right) + 2 \quad \forall i_1'$$

$$1 \leq n_{i_1}^{(1)} \leq 5 \quad \forall i$$

$$10 \leq \sum_{i_1} n_{i_1}^{(1)} \leq 15$$

1.2 Adjacency matrix constraints. (Slide 6)
1.3 Second-order group constraints. (Slide 6)

2. Pure component property constraints:

$$lco + \sum_{i_1} n_{i_1}^{(1)} lc50_{i_1}^{(1)} + \sum_{i_2} n_{i_2}^{(2)} lc50_{i_2}^{(2)} \geq 3.16$$

$$Spo + \sum_{i_1} n_{i_1}^{(1)} Sp_{i_1}^{(1)} + \sum_{i_2} n_{i_2}^{(2)} Sp_{i_2}^{(2)} \leq 25$$

$$\sum_{i_1} n_{i_1}^{(1)} clp_{i_1}^{(1)} \geq 343.15^2$$

$$Fpo + \sum_{i_1} ni_1^{(1)} Fp_{i_1}^{(1)} + \sum_{i_2} n_{i_2}^{(2)} Fp_{i_2}^{(2)} \geq 343.15$$

$$0.1 \leq Vmo + \sum_{i_1} ni_1^{(1)} Vm_{i_1}^{(1)} + \sum_{i_2} n_{i_2}^{(2)} Vm_{i_2}^{(2)} \leq 0.3$$

- #Equations: 1,324,764; #Variables: 1,286,278 (1,286,266 Integer var.).
- Step 4: This model can be solved directly using GAMS CPLEX solver.

15AIChE
Annual Meeting, Salt Lake City, UT

DTU

Surfactant Design (3/3)

Table: Optimization results of surfactant design problem

n_1	1 CH_3, 5 CH_2, 1 CH, 4 aCH, 1 aC−OH, 1 CH_2COO, 1 aC−O, 1 OCH_2CH_2OH
n_2	1 AROMRINGs1s4
$lc50$	4.124log(mol/L)
Sp	24.732MPa$^{0.5}$
clp	300.016K
Fp	552.383K
Vm	0.281cc/mol

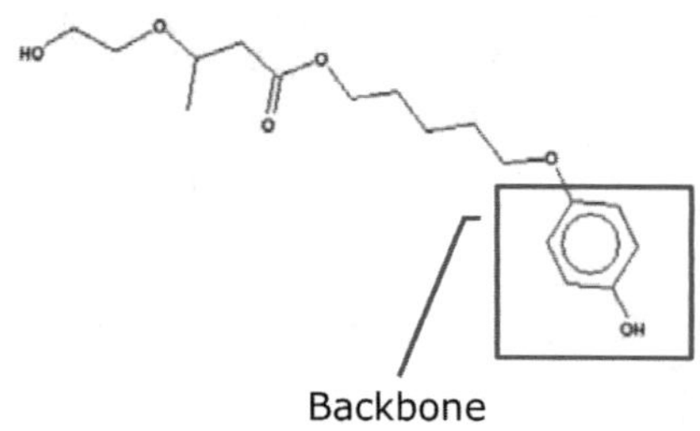

15AIChE
Annual Meeting, Salt Lake City, UT

Refrigerant Design - Product-Process Design (1/3)

- Step 1: Design a new refrigerant for the replacement of R-134a, the new fluid should have (1) similar/better properties and thermodynamic behaviour; (2) reduced environmental impact.
- Step 2:
  - **Objective:** Coefficient of performance (COP) should be maximized.
  - **Group selection:** Fluids can contain C, H, F and Cl. 30 molecular groups are selected from 220 groups (Marrero and Gani, (2001)).
  - **Target properties for product and process:**

Target Property	Lower bound	Upper bound
Molar mass	-	110g/mol
Normal boiling point	-	250K
Critical temperature	350K	400K
Critical pressure	30bar	50bar
Thermal conductivity (liquid, Tb)	0.08W/m K	-
Enthalpy of Vaporization at Tb	200KJ/Kg	-
ODP	-	0
GWP	-	1400
Atmospheric Lifetime *	-	14years
Process needs	**Lower bound**	**Upper bound**
Evaporation temperature	313.15K	333.15K
Condensation temperature	265.65K	280.15K

15AIChE
Annual Meeting, Salt Lake City, UT

Refrigerant Design (2/3)

$$\max COP = \frac{H_1(Z_{V1},T_1,P1) - H_3(Z_{L3},T_3,P_3)}{H_2(Z_{V2},T_2,P_2) - H_1(Z_{V1},T_1,P_1)}$$

s.t.

1. Structural constraints.
2. Property constraints.
3. Process constraints:

$$\ln \psi_{V1} = \ln \psi_{L1}$$
$$\ln \psi_{V2} = \ln \psi_{L2}$$
$$H_{L3}(T_3,P_3) = H_{L_4}(T_4,P_4)$$
$$S_{V2}(T_2,P_2) = S_{V1}(T_1,P_1)$$
$$P_1 = P4 \quad P_2 = P_3 \quad T_1 = T_4$$
$$P_2 \leq 0.8P_c \quad T_2 \leq 0.8T_c$$
$$265.65K \leq T_1 \leq 280.15K$$
$$313.15K \leq T_3 \leq 333.15K$$
$$Z_{V1}^3 - Z_{V1}^2 + Z_{V1}(-B^2 + A - B) - AB = 0$$
$$Z_{V2}^3 - Z_{V2}^2 + Z_{V2}(-B^2 + A - B) - AB = 0$$
$$Z_{V3}^3 - Z_{V3}^2 + Z_{V3}(-B^2 + A - B) - AB = 0$$

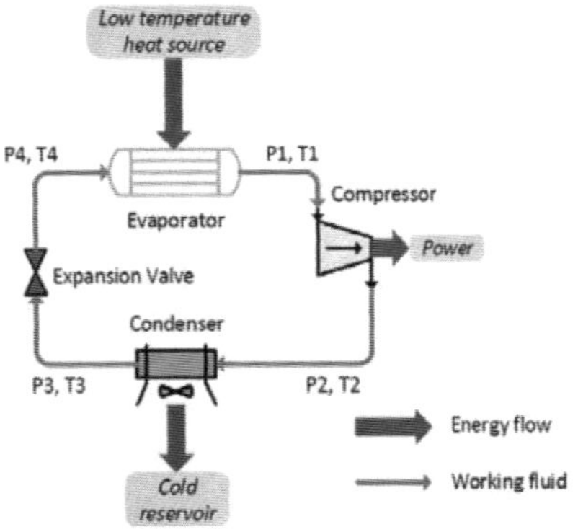

Constraints to ensure correct cubic EOS root selection (Kamath et al., 2010):

$$f(Z_{V1}) = 0 \quad f'(Z_{V1}) \geq 0 \quad f''(Z_{V1}) \geq 0$$
$$f(Z_{V2}) = 0 \quad f'(Z_{V2}) \geq 0 \quad f''(Z_{V2}) \geq 0$$
$$f(Z_{V3}) = 0 \quad f'(Z_{V3}) \geq 0 \quad f''(Z_{V3}) \geq 0$$

[1]Stefano Cignitti, et al., CAST Division Plenary

Refrigerant Design (3/3)

Table: Optimization results for refrigerant design problem

	r134a	r152a	3,3,3-diFluoropropyne	3,3-diFluoropropene
	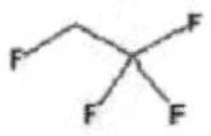	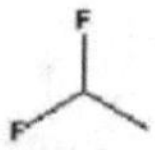		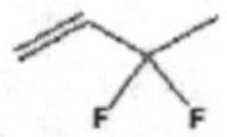
Mw (g/mol)	66	102	94	92
Tb (K)	230	236	242	246
Tc (K)	366	374	369	392
ω	0.27	0.33	0.28	0.14
k (W/m K)	0.097	0.082	0.086	0.093
ΔH_v at Tb (J/mol)	28562	26523	18635	21312
ODP	0	0	0	0
GWP	**1370**	**133**	**134.6**	**2.4**
Atmospheric lifetime (years)	**13.4**	**1.5**	-	-
COP	**10.1**	**12.8**	**9.5**	**11.0**

Die Attach Adhesive (DAA) Design - Product Design with Economic Analysis (1/3)

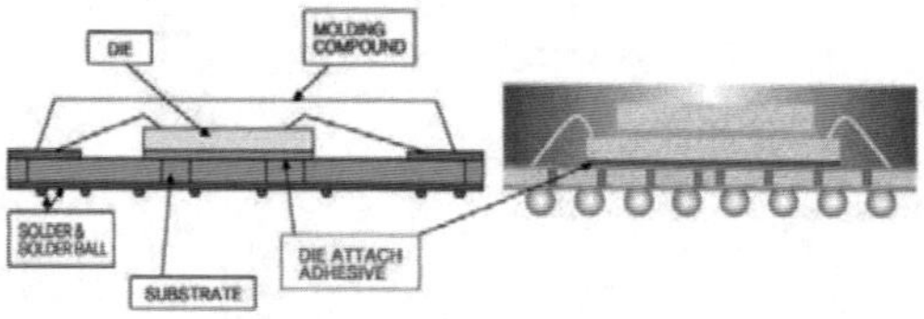

- Step 1: Die attach adhesive design used in LED lamp. It is crucial to remove the heat from the chip as fast as possible (high thermal conductivity)
- Step 2:
  - **Objective:** Net present value (NPV) should be maximized.
  - **Ingredients:** Polymer (designed), microparticle (Ag or Cu) and nanoparticle (Ag or Cu).
  - **Group selection for polymer:** -C6H4-, -O-, CH2, CH, C=O, -S-, -SO2-, from 100 groups (D.W. van Krevelen, 2009)
  - **Property constraints:** The thermal conductivity of the product should equal 30 W/m K

*Collaboration with HKUST (Kelvin Zhang and Prof. Ka M. Ng)

15AIChE
Annual Meeting, Salt Lake City, UT

DTU

Die Attach Adhesive Design (2/3)

$$\max npv = \sum_{j=-n}^{m} \frac{pcf_j}{(1+R)^j}$$

s.t.

1. Structural constraints.
2. Property constraints (Group contribution method).
3. Economic model:

$$pcf_j = \begin{cases} -TDC_j - TCI_j - \Delta NWC_j & j \leq 0 \\ (p_j d_j - c_{m,j} - Adv_j - FC_j)(1-t) + tD_j + \Delta NWC_j & j > 0 \end{cases}$$

$$\begin{aligned} c_{m,j} &= (w_{mp} c_{mp} + w_{np} c_{np} + c_{m,pc}) d_j \quad j \geq 1 \\ p_j d_j &= \left(\frac{\alpha_j}{\beta}\right)^{\rho} P_j^C \left(\frac{Y - p_j d_j}{P_j^C}\right)^{1-\rho} d_j^{\rho} \quad j \geq 1 \\ Y &= p_j d_j + P_j^C d_j^C \quad j \geq 1 \end{aligned}$$

15 DTU Chemical Engineering, Technical University of Denmark

17 December 2015

15AIChE
Annual Meeting, Salt Lake City, UT

DTU

Die Attach Adhesive Design (3/3)

Optimization results of die attach adhesive design problem

	Polymer			Microparticles			Nanoparticles	
Chemical	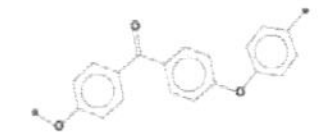			Cu			Cu	
Weight fraction	0.1187			0.7974			0.0839	
NPV	Net present value (MM$)						80.1024	
j	Years	-1	0	1	2	3	4	5
PCF_j	Project cash flow (MM$)	-1	-2.5	27.8867	27.9587	27.8819	27.8358	28.8358
p_j	Selling price ($/Kg)	-	-	1351	1351	1351	1351	1351
d_j	Demand (t)	-	-	55.4647	55.4647	55.4647	55.4647	55.4647
c_j	Product cost (MM$)	-	-	27.654	27.654	27.654	27.654	27.654
Adv_j	Advertisement cost (MM$)	-	-	0.5	0.5	0.5	0.5	0.5
$d_{c,j}$	Competitor's demand (t)	-	-	8.3561	8.3561	8.3561	8.3561	8.3561

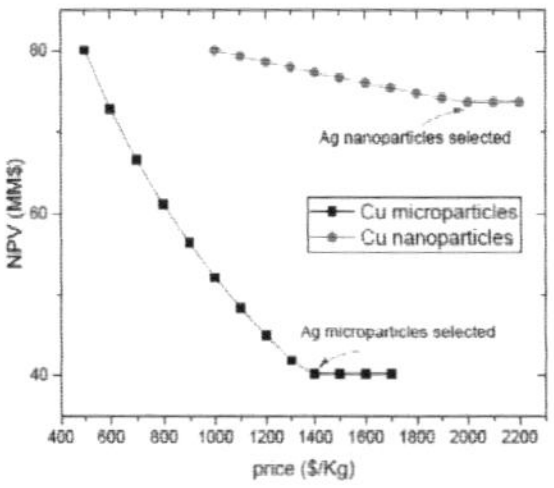

Metal particles price vs NPV

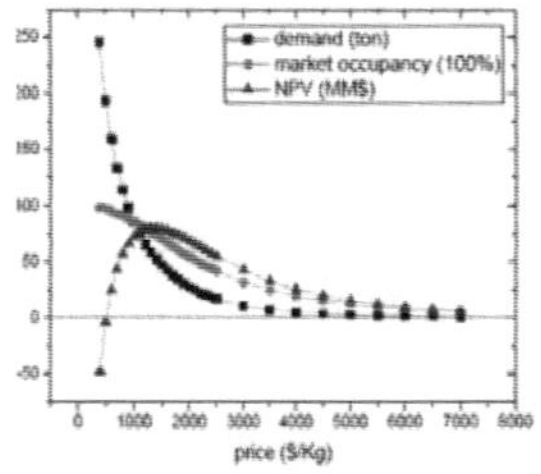

Product price vs demand, market occupancy and NPV

16 DTU Chemical Engineering, Technical University of Denmark

17 December 2015

Conclusion

- A computer-aided framework was developed in this work.
- This framework has been shown to systematically define and solve a large MINLP product design, product-process design, and product-process design with economic analysis problems.
- Generate and test methods can only partially solve these problems.
- Higher order description of groups and molecular structure prediction is integrated in this new MINLP mathematical programming model.
- This model is tested using a large range of product design problems (surfactants, refrigerants and polymers) and for consideration of sustainability inclusion through economic and environmental models.

427253 Designing Optimal Mixtures Using Generalized Disjunctive Programming: Convex Hull Reformulation

Abstract:

The design of mixtures is a topic of great importance across many industrial sectors, as part of process development and of product design. Computer-Aided Mixture/blend Design (CAM^{b}D) [1] has emerged as an attractive methodology which has been successfully applied to designing various types of mixtures. Within the CAM^{b}D framework, these design problems have often been posed as Mixed-Integer Nonlinear Programming (MINLP) problems, where the optimal molecular structure is determined with respect to a set of property constraints. In most existing methodologies a reduced version of the general CAM^{b}D problem has been addressed: the number of mixture ingredients is fixed (usually a binary mixture is considered) and the identity of a compound (or of all compounds) that can participate in the mixture is chosen from a given set of candidate molecules. Solving the resulting optimization problems can be quite challenging, as these are combinatorial in nature due to the presence of binary variables and to the highly nonlinear expressions that relate composition, structure and physical properties.

In working towards a generalised CAM^{b}D problem, a novel methodology where the number, identity and compositions of mixture constituents are optimised simultaneously, was recently proposed [2,3]. In order to address the difficulties arising from the complexity of expressing the problem within a mathematical framework, a logic-based methodology, Generalized Disjunctive Programming (GDP) [4], was used to formulate the discrete choices of a mixture design problem. The design methodology was first developed for a standard formulation with a fixed number of components (restricted problem) and then generalised to allow the number of mixture ingredients to vary up to a maximum number (general problem). In this approach, the designed components in the mixture are selected from a list of candidate pure compounds. The GDP problems are formulated as MINLP problems in order to exploit the available algorithms for solving optimisation problems. The formulations are converted into MINLP problems by replacing the Boolean variables with binary ones and converting the logic propositions into linear inequalities. In this initial work [2,3], conditional constraints inside the disjunctions were formulated using the Big-M (BM) approach [4]. The BM approach is the simplest representation of a GDP problem in a mixed-integer form. However, it introduces a large parameter (Big-M) which in most cases is specified intuitively and can yield to poor relaxations. Here, we extend this work by converting the GDP models into MINLP problems using the Convex Hull (CH) reformulation [5]. The CH reformulation introduces a new set of disaggregated variables and new constraints, increasing the size of the problem. It can however be shown [6] that when the discrete domain of the CH formulation is relaxed, it gives at least as tight or tighter relaxations than BM.

The design methodology is applied to two case studies: the first one aimed to design an optimal solvent mixture to dissolve ibuprofen and the second one to design the most effective solvent mixture to extract acetic acid from water. The results of the problems studied show that the proposed formulations offer a promising approach to the problem of mixture design, as the simultaneous design of the optimal number, identity and compositions of the components that participate in a mixture, is achieved. The CH reformulation is compared to the Big-M approach and both methods appear suitable for formulating mixture problems effectively. Finally, better performance is obtained with a solvent mixture in both examples, showing the benefit of using mixtures instead of pure solvents to attain enhanced properties.

References

1. L.E.K. Achenie, R. Gani, and V. Venkatasubramanian. *Computer aided molecular design: theory and practice*. Elsevier, Amsterdam, Netherlands, 2003.
2. P.T. Akula, P.M. Kleniati, and C.S. Adjiman. On the design of optimal solvent mixtures using generalised disjunctive programming. In *22nd European Symposium on Computer Aided Process Engineering*, pages 1-5, 2012.
3. S. Jonuzaj, P.T. Akula, P.M. Kleniati, and C.S. Adjiman. The formulation of optimal mixtures using generalized disjunctive programming: choosing components from a pre-defined list. *Journal paper in preparation*, 2015.
4. R. Raman and I.E. Grossmann. Modelling and computational techniques for logic based integer programming. *Computers & Chemical Engineering*, 18(7):563-578, 1994.
5. S. Lee and I.E. Grossmann. New algorithms for nonlinear generalized disjunctive programming. *Computers & Chemical Engineering*, 24(9-10):2125-2141, 2000.
6. S. Lee and I.E. Grossmann. Generalized convex disjunctive programming: convex hull relaxation. *Computational Optimization and Applications*, 26:83-100, 2003.

432235 A COSMO-Based Approach to Computer-Aided Mixture Design

Abstract:

A relatively well-known field in the literature, Computer-Aided Molecular Design (CAMD) defines a class of problems aimed at designing a single, optimal molecular structure to best fit some purpose. While CAMD[1] has seen some success in the literature, its mixture analogue, Computer-Aided Mixture Design (CAMxD), still has some significant limitations. CAMxD is, unsurprisingly, much more complicated. This class of problems now involves the design of a blend of compounds – both every compound's molecular structure and mole fraction in the composition – and this greatly increases the number of variables in the problem and adds many more nonlinear constraints in the form phase equilibria equations, miscibility checks, and safeguards against including two molecular species which might react.

The traditional approach to CAMxD in the literature has two main shortcomings: (1) many approaches rely on enumerating every possible structure for every possible component and then checking every combination for mixture feasibility; and (2) the search space is inherently curtailed because these approaches often solve the mixture part of the problem with UNIFAC[2] method, which can only work when a regressed interaction parameter is known for every pair of molecular substructures in the blend. These issues have resulted in approaches to the CAMxD problem that can only investigate a very limited area of the molecular search space.

Our approach to the CAMxD problem circumvents these two issues by: (1) using a derivative-free optimization scheme to design for structures and test their mixture properties on-the-fly; and (2) deriving a group-contribution method to estimate COSMO-RS[3] and COSMO-SAC[4] sigma profiles, which can be used to calculate mixture properties without the need for binary interaction parameters. These two steps enable a much more efficient optimization of a much larger chemical search space. In this presentation, we will outline our approach in more detail and discuss how the COSMO group contribution methods were made. We will also demonstrate our algorithm as applied to a solvent design problem and a reaction rates optimization problem.

[1] Group-contribution based estimation of pure component properties. Jorge Marrero and Rafiqul Gani.*Fluid Phase Equilibria.* 2001. 183, 183-208.

[2] Group-contribution estimation of activity coefficients in nonideal liquid mixtures. Aage Fredenslund, Russel L. Jones, and John M. Prausnitz. *AIChE Journal*. 1975. 21(6), 1086-99.

[3] Fast solvent screening via quantum chemistry: COSMO RS approach. Frank Eckert and Andreas Klamt. " *AIChE Journal.* 2002. 48(2), 369-385.

[4] A priori phase equilibrium prediction from a segment contribution solvation model. Shiang-Tai Lin and Stanley I. Sandler. " *Industrial & engineering chemistry research.* 2002. 41(5), 899-913.

(173f) Coupling Experiments and Computational Tools to Estimate Thermal Properties for Product Design in the Fat-Based Food Industry

Teles dos Santos, M., Universidade de São Paulo - USP
Viana, I. S., Universidade de São Paulo
Ract, J. N. R., Universidade de São Paulo - USP
Le Roux, G. A. C., Polytechnic School of the University of São Paulo
Originally presented on: 11/9/2015 14:15:00 - 14:36:00

Product design has been evaluated as one of the challenges that chemical engineering has to face to help industry meet market demands for value-added products with higher performance [1]. In the food and pharmaceutical industry, the design of fat-based products is of great importance, since many foods have their desired properties related to a complex phase behavior involving lipids and many cosmetic and pharmacological products use vegetable oils as a lipidic medium in which active principles and fine chemicals are dispersed. Attempting to use Process Systems Engineering tools (namely thermodynamic modeling and computational/numerical approach), for product design, the present work coupled experimental studies and computational simulations to access the thermal properties of different scales of triacylglycerol mixtures, ranging from simple binary mixtures to ternary blends of vegetables oils.

In order to test the developed tools in raw materials used by industry, different blends formed by palm stearin (PS), fully hydrogenated soybean oil (FHSO) and canola oil were studied. The solid fat content at different temperatures, melting points and changes in heat capacity due to solid-liquid transitions are covered by the present work, both experimentally and computationally. Computational simulations and experimental results are compared and discussed. The following experiments were conducted: fatty acids composition identification, stereo distribution of fatty acids in the glycerol structure, softening point and thermal analysis (differential scanning calorimetry). The computational simulations were performed by the following steps: prediction of triacylglycerol composition of the blends by using statistical methods (random and non-random distribution of fatty acids in the glycerol structure); predictions of the distribution of molecules among solid and liquid phases by direct minimization of Gibbs free energy function, using a Generalized Reduced Gradient Method (CONOPT 3 GAMS solver). The optimization algorithm was then coupled with the main program written in FORTRAN 90, which handles the calculation of interaction parameters, melting temperature and melting enthalpy and the generation of triacylglycerols from fatty acids data [2]. The Solid-liquid equilibrium problem was then solved over a temperature range to be compared with experimental data in different temperatures and at different blend ratios (single vegetable oil, binary blends and ternary blends).

As the demand for more sophisticated fat products increases and the simple physical mixture (blending) is not sufficient for matching the desired properties, new technologies have emerged. The chemical interesterification reaction is one important technique that changes the triacylglycerol profile of the blends and, thus, modifies their physical properties. This reaction uses catalysts (e.g. sodium methoxide) and heat to promote a random distribution of fatty acids among the three positions of glycerol [3]. In order to predict how the reaction affects the final solid fat content of the blends, this work simulates the effect of the reaction by using random distribution of fatty acid in the glycerol chain and computing the melting profile of the reaction product.

REFERENCES

[1] Hill, M. (2009). "Chemical product engineering — the third paradigm." *Comput. Chem. Eng.*33, 947–953.

[2] Teles dos Santos, M.; Gerbaud, V.; le Roux, G.A.C. (2013). "Modeling and simulation of melting curves and chemical interesterification of binary blends of vegetable oils." Chem. Eng. Sci. 87, 14–22.

[3] Rodrigues-Ract, J.N.; Cotting, L.N.; Poltronieri, T.P.; da Silva, R.C.; Gioielli, L.A. (2010). "Crystallization behavior of structured lipids by chemical interesterification of milkfat and sunflower oil." Food Sci. Techn. 30, 258-267.

433479 Multi Objective Molecular Design of Reactants and Products

Abstract:

In reactive systems, computer-aided molecular design (CAMD) is performed to design solvents that enhance reaction kinetics, improve catalyst activity/lifetime as well as reactants and products. CAMD of reactants and products helps in their selection such that target properties of interest are optimized with satisfaction of property constraints. Owing to the scarcity of CAMD methodologies that design reactants and products, there has been an increased interest to address this issue. The few recently available methodologies to design reactants and products however have their shortcomings. The earliest available methodologies are restricted to cases where only structures of a single reactant and product are variable. Also, employment of linear property models and lack of a methodology to treat property constraints have been some of the other restrictions. Dev et al. [1, 2] recently developed two algorithms to design reactants and products which are unrestricted by the number of reactants and products whose structures are uknown. They also made provisions for non-linear property models and property constraints. In their first algorithm [1], only the dominant properties of the products are optimized. Also, only the products are subjected to their individual set of property constraints. In their second algorithm [2], properties that are dependent on the structures of both reactants and products are optimized. Also, all reactants and products are subjected to their respective set of property constraints. In this work, we have developed an algorithm that generates structures of reactants and products irrespective of their numbers such that each reactant and product's target property(s) is to be optimized. Also, each reactant and product is subjected to a set of property constraints. All properties have been expressed in terms of molecular structure using different types of property models. Since there exists a need to treat different types of property models on a single platform, signature descriptors have been utilized. A variety of property models can be expressed in terms of signature descriptors which are molecular building blocks. They enable the simultaneous utilization of Quantitative Structure Activity/Property Relationships (QSARs/QSPRs) and Group Contribution Models (GCMs) [3]. In the formulated problem, all property models are expressed in terms of the occurrence number of signature descriptors. Relationships have been developed that relate the occurrence number of signatures of reactants and products. This is achieved by tracking the atoms in the reactants to the products by taking into consideration the reaction mechanism. Thus the reactants and products have been structurally related. Due to this structural relation, property constraints acting on reactants and products influence the selection of both optimal reactants and products. The objective functions of each of the reactants and products display a similar influence. Since the selection of each reactant and product is dependent on the property constraints and property objective functions of the other reactants and products in addition to its own, the problem is formulated as a multi-objective optimization problem when objectives are conflicting. This problem is solved using the Augmented Epsilon-Constraint Method (AUGMECON) developed by Mavrotas [4]. AUGMECON generates pareto optimal solutions while completely avoiding weakly pareto optimal solutions. The generated solutions also include unsupported pareto optimal solutions. Specifically for multi-objective integer programming (MOIP) problems, an improved version of AUGMECON, AUGMECON2 [5] can also be utilized. AUGMECON2 is very efficient in generating the entire pareto optimal set for MOIP problems compared to AUGMECON and some earlier available methods. In this contribution, the utility and novelty of the developed methodology has been highlighted through a case study.

[1] V.A. Dev, N.G. Chemmangattuvalappil, M.R. Eden, 2014, Computer Aided Chemical Engineering, 34, 291-296

[2] V.A. Dev, N.G. Chemmangattuvalappil, M.R. Eden, 2015, Computer Aided Chemical Engineering, 37, 1445-1450

[3] N.G. Chemmangattuvalappil, C.C. Solvason, S. Bommareddy, M.R. Eden, 2010, Computers & Chemical Engineering, 34, 12, 2062-2071

[4] G. Mavrotas, 2009, Applied Mathematics and Computation, 213, 455-465

[5] G. Mavrotas, K. Florios, 2013, Applied Mathematics and Computation, 219, 9652-9669

437575 Cheml- a Machine Learning and Informatics Program Suite for the Chemical and Materials Sciences

Abstract:

Computational chemistry and materials science are useful assets to study the properties of novel compounds, materials, and reactions. While these investigations are often valuable for the specific systems at hand, their broader impact tends to be more limited. Our research program focuses on merging computational chemistry and materials modeling with Big Data ideas. The virtual high-throughput screening of candidate libraries allows us to generate huge quantum chemical data sets, which can be employed for data-driven discovery. To extract an understanding of the underlying structure-property relationships from these data sets we develop chemical data mining tools, which are assembled in our CheML software suite. These tools state-of-the-art techniques from machine learning, statistical learning, and informatics. We will discuss their utility and the resulting models for the prediction of molecular properties without the need for expensive quantum chemical calculations.

430275 A Molecular Workbench for Computational Soft Materials

Abstract:

The objective of the Materials Genome Initiative (MGI) is a new R&D paradigm enabling accelerated materials design via computation leading to a shortening of the time and cost needed to bring new materials to market. This new paradigm will be fostered through the creation of national infrastructure for materials data sharing and analysis that supports Integrated Computational Materials Engineering (ICME), an emerging multiscale paradigm for the integrated modeling of materials from the quantum or molecular level, upward in both length and time scale to the continuum scale represented by structural components. An essential element to meeting these objectives for polymers and related soft materials is the development of a coherent and progressive framework for carrying out coarse-grained modeling at mesoscopic length scales between the atomistic and continuum. Such a framework requires meeting the theoretical challenge of describing the representation and interoperability of materials data and the interoperation of modeling systems at multiple length and time scales, as well as the practical challenge of creating a computing environment that realizes its instantiation.

In this seminar, we describe a computational workbench being developed to provide an environment for the integration of coarse-graining algorithms into a hierarchical, multiscale, modeling platform. In the application, the first level of description is atomistic data which contains the highest level of information about the material. Data and models at progressively higher scales are integrated into the environment by taking advantage of the hierarchical and modular environment. The platform is hierarchical in the sense that the native data structure supports a multiscale description of a material starting with atoms through progressively higher levels of coarse-graining. The environment is modular in the sense that it supports the addition of new functionality through Python scripting and compiled C++ libraries as run-time plugins. Together, these features enable algorithms for different levels of coarse-graining to be integrated into the same environment and function together, given proper theoretical basis. The final element of the application is that of reference data. Components of the NIST Materials Data Curator are being integrated into the environment to support ontology based database descriptions allowing users to build progressive, materials reference libraries.

References

1. Materials Data and Informatics, http://www.nist.gov/itl/ssd/is/materials-data-and-informatics.cfm

432951 Mdn: A Web Portal for Network Analysis of Molecular Dynamics Simulations

Abstract:

We introduce a web portal that employs network theory for the analysis of trajectories from molecular dynamics simulations. Users can create protein energy networks following methodology previously introduced by our group, and identify residues that are important for signal propagation, as well as measure the efficiency of signal propagation by calculating the network coupling. This new tool, called MDN, was used to characterize signal propagation in E. coli heat shock protein 70-kDa (DnaK). Two DnaK variants, experimentally shown to be allosterically active, exhibit higher network coupling than two inactive variants. In addition, calculations of partial coupling suggest that this quantity could be used as part of the criteria to determine 'pocket druggability' in drug discovery studies.

430786 Development of the Parallel Gibbs Ensemble Monte Carlo Simulation Engine Gomc

Abstract:

GPU Optimized Monte Carlo (GOMC) is a high performance, open source, Gibbs Ensemble Monte Carlo simulation engine that is capable of simulating complex molecular systems containing over 100,000 atoms with performance 10-100 times faster than existing serial codes. GOMC uses graphics processing units (GPUs) for the calculation of pairwise energies via energetic decomposition. The code supports simulations in the canonical, isobaric-isothermal, grand canonical and Gibbs ensembles (NVT and NPT) for molecules of arbitrary geometry. GOMC utilizes CHARMM style input files (topology, parameter, PSF, PDB), allowing the use of a single set of input files for CHARMM, NAMD[1] or GOMC. Output from GOMC is compatible with VMD[2].

In this talk, we discuss the implementation of new functionality, which includes support for the calculation of electrostatic energies, new potential functions for non-bonded interactions, and coarse-grained force fields, such as MARTINI. Optimization strategies unique to simulations with changing numbers of interaction sites, e.g. Gibbs Ensemble and grand canonical Monte Carlo are discussed. Performance data are presented for serial, OpenMP and GPU versions of the code.

1. Phillips, J.C., et al., *Scalable molecular dynamics with NAMD.* Journal of Computational Chemistry, 2005. **26**(16): p. 1781-1802.

2. Humphrey, W., A. Dalke, and K. Schulten, *VMD: Visual molecular dynamics.* Journal of Molecular Graphics & Modelling, 1996. **14**(1): p. 33-38.

437565 Scalable Forward Flux Sampling, Scaffs: Enabling Large Scale Simulations of Rare Events

Abstract:

Forward flux sampling (FFS) is an established scientific method for sampling rare events in simulations. However, as the difficulty of the scientific problem increases, the amount of data and the number of tasks required for FFS is challenging to manage with traditional scripting tools and languages for high performance computing. The Scalable FFS (ScaFFS) software framework has been developed to address these challenges. ScaFFS utilizes Hadoop to manage a large number of tasks and data for large-scale FFS simulations. The framework is shown to be highly scalable and is able to the support large-scale FFS simulations necessary to extend the use of FFS to complex molecular systems on commodity cluster computing systems. We will present results of the performance of ScaFFS and results obtained using ScaFFS related to ice and gas hydrate nucleation.

413843 Oxidative Coupling of Methane: Descriptors for Activity and Selectivity

Abstract:

Oxidative Coupling of Methane: descriptors for activity and selectivity

Gaurav Kumar, Sai Lap Jacky Lau and Michael J. Janik*

Department of Chemical Engineering, Pennsylvania State University, University Park, PA

*mjanik@engr.psu.edu

Increasing natural gas reserves in the recent years motivate further development of processes to convert methane into value-added products. Since the initial description [1] of catalytic oxidative coupling of methane (OCM) to form C_2 hydrocarbons, various pure/doped metal-oxide catalysts have been examined for their methane conversion and C_2 selectivity. This ongoing search to find high performing catalysts has been pre-dominantly empirical, and the relationship between active site properties and catalyst activity and selectivity remains unclear. A known dependence of the OCM activity and selectivity on material property(s) enables us to tune active site properties to optimize catalyst performance.

This study uses density functional theory (DFT) methods and the established OCM reaction mechanism [2] to relate activity and selectivity to the surface electronic and structural properties of the catalyst. C-H activation and °¤CH_3 radical adsorption determine the activity and selectivity of a catalyst, respectively. Since both of these elementary processes reduce the catalyst surface, *we hypothesize that the activity (measured by C-H activation) and selectivity* *(measured by °¤CH_3 adsorption) of the catalyst correlate with surface reducibility.* C-H activation energy and °¤CH_3 adsorption energy have been plotted against the oxygen vacancy formation energy ¥ÄE_{vac} of various metal-doped CeO_2, doped MgO, doped TiO_2, ZnO and TbO_x. Computational results (Figure 1) show a linear relation of the vacancy formation energy (surface reducibility), with the C-H activation energy and °¤CH_3 adsorption energy.

Ceria has a lower C-H activation energy which suggests that it is highly active and would offer better methane conversion. However, it also binds °¤CH_3 strongly leading to over-oxidation of methane and thus gives low C_2 selectivity. Conversely, high C-H activation energy and weak °¤CH_3 adsorption imply that pure/doped MgO catalysts will have better C_2 selectivity but low activity. Therefore, a trade-off between activity and selectivity is inherent in the active site. Sub-correlations to that shown in Figure 1 demonstrate that this correlation holds across a single material with increasing state of reduction as well as with change in U parameter used in the DFT+U method. Methods on finding an optimal trade-off in activity and selectivity, as well as potential approaches to breaking this correlation, will be discussed.

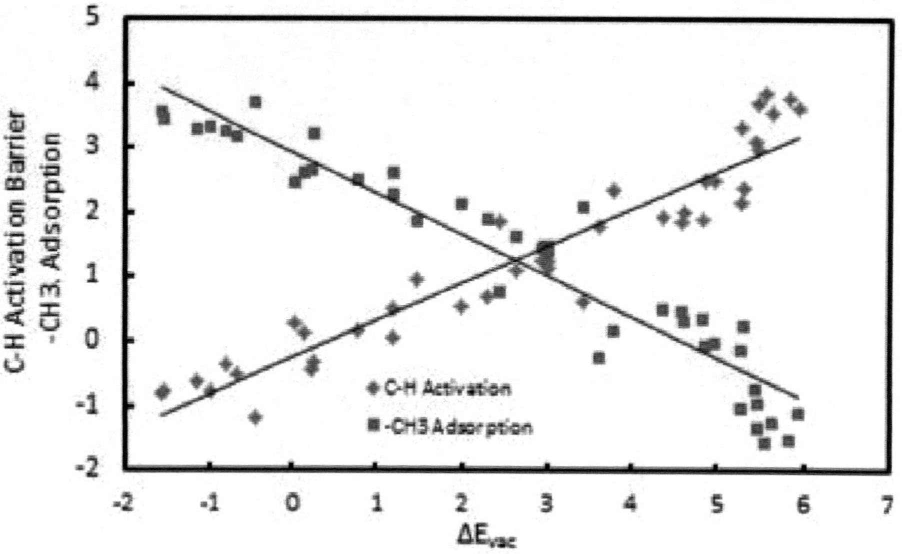

Figure 1. Correlation between surface reducibility and activity/selectivity

References

1. G. E. Keller, M. M. Bhasin, *Journal of Catalysis* 73 (1982) 9.

2. T. Ito, J.H. Lunsford, *Nature* 314 (1985) 721.

416770 Entropic and Enthalpic Driving Forces on Morphology in Polymer Grafted Particle Filled Nanocomposites: Integral Equation Theory and Molecular Simulations

Abstract:

Polymer nanocomposites are a class of materials that consist of a polymer matrix embedded with nanoscale fillers or additives that enhance the inherent properties of the matrix polymer. To engineer polymer nanocomposites for specific applications with target macroscopic properties (e.g. photovoltaics, optics, photonics, parts of automobiles, etc.) it is important to have design rules that relate molecular features to morphologies of the composite. Using theory and simulation, we studied the effects of tuning the properties of grafted and matrix chains in polymer grafted nanoparticle filled composites in order to tune the entropic and enthalpic driving forces for particle dispersion and aggregation. Initially, we studied polymer nanocomposites with homopolymer grafted particles in a homopolymer matrix, at high grafting density and with *chemically identical* graft and matrix polymers. This allowed us to probe how entropic driving forces affect nanocomposite morphologies. We found that increasing the polydispersity in grafted chain lengths or decreasing the graft and matrix chain flexibility stabilizes the dispersed phase of polymer nanocomposites due to increased wetting of the grafted layer by matrix chains. We further showed that this increased wetting of the grafted layer was primarily driven by increased graft-matrix mixing entropy and decreased conformational entropy penalty of the matrix wetting the grafted layers. In more recent work, we explored composites with *chemically different* graft and matrix polymers which have, in addition to the entropic driving forces described above, enthalpic driving forces that drive the system towards maximizing attractive graft-graft, matrix-matrix, and graft-matrix interactions. We found that the sharp particle dispersion-aggregation transition appears to be distinct from the continuous wetting-dewetting transition, which is in direct contrast to *chemically identical* graft/matrix systems where dispersion-aggregation and wetting-dewetting are treated synonymously. In this poster I will describe the above theoretical and computational results, along with experimental results that have confirmed our predictions.

417327 Hybrid Atomistic and Coarse-Grained Molecular Dynamics Simulations of Polyethylene Glycol (PEG) Chains in Explicit Water for Designing Peg Based Biomaterials

Abstract:

Molecular dynamics (MD) is a well established technique to simulate the structure and dynamics of soft matter in general, and of biomolecular systems in particular. While simulations at the all-atom (AA) or united-atom (UA) level maintain atomistic chemical details, they are computationally intensive. On the other hand coarse-grained (CG) models, in which several atoms are grouped together into effective interaction sites (CG beads), have been proven successful in achieving large length and time scales, while sacrificing atomistic level details. The *in-silico* design of polymeric biomaterials requires MD simulations that retain chemical detail for the essential biofunctional species and achieve large length and time scales pertinent to non-functional polymers. We are thus motivated to use hybrid atomistic – CG simulations, in which part of the system that is critical to the function of the biomaterial is represented at atomic resolution and the remaining part of the biomaterial that hosts the functional molecule is represented in coarse-grained level. Calculating the full atomistically detailed dynamics only in important regions of the system and reducing the complexity for the rest of the biomaterial can significantly reduce the computational cost and allow for optimal design of experimentally relevant length scale polymer based biomaterials. In this poster, we will present our recent efforts in building a hybrid simulation approach for poly(ethylene glycol) or PEG based biomaterials in aqueous medium for drug-delivery applications.

419477 A New Lattice Monte-Carlo Simulation for the Dielectric Inhomogeneity of Ion-Containing Liquids

Abstract:

We develop a new, rapid lattice Monte-Carlo simulation of ion-containing liquids to capture the reorganization of solvent dipoles under external electrostatic fields. Our simulation accounts for the effects of saturated dipoles near ions on the angstrom scale and hence spatial variations in the dielectric function. Our results are in reasonable agreement with the analytical solutions to the dielectric continuum theory of Booth for single ions, ion pairs, and ionic cross-links. We will also show the substantial disparity between the dielectric functions for like and unlike charges on the nanometer scale. Importantly, a contacting cation-anion pair cannot be literally taken as ``charge-neutral species" in terms of the solvation energy. Moreover, even when the two charges are separated by 1 [nm], a significant correlation in the dielectric function may arise. Our simulation also provides the dependence of the bulk dielectric value of water on ion concentrations, which is consistent with experimental data.

419959 Theoretical Investigation of Thermal Oxidation of Carbon-Coated Aluminum Nanoparticles Using the Reaxff Reactive Force Field

Abstract:

The thermal oxidation of aluminum nanoparticles (ANPs) has been widely investigated for solid-fuel rockets due to their high enthalpy of oxidation during the combustion process. Essentially, ANPs have an accelerated reaction rate because of an increased surface area-to-volume ratio as compared to larger particles (e.g., a micron size), making them attractive in reactive additives for a solid propellant. Unfortunately, the increased reaction rate of ANPs also leads to a critical issue where excessive oxide layers on bare ANPs can be formed and grown prior to the combustion process. As a result, the energy release per unit mass during thermal oxidation significantly decreases because the naturally-formed oxide layer serves as a dead weight. In order to overcome this challenge, a carbon coating on bare ANPs has been proposed because of the several following advantages: (a) carbon is non-reactive at low temperature, but reactive at elevated temperatures; (b) carbon-coated ANPs have a hydrophobic nature; and (c) carbon can serve as an additional fuel-constituent. However, growth kinetics of the oxide layer on carbon-coated ANPs are not yet fully understood. In addition, a direct comparison of surface chemistry in bare and carbon-coated ANPs has not been done. In this work, we investigate the surface chemistry of the thermal oxidation of carbon-coated ANPs as compared to bare ANPs using the ReaxFF reactive force field. For this purpose, a ReaxFF description for Al/C interactions was developed, and then ReaxFF-molecular dynamics (MD) simulations were performed to investigate the growth kinetics of the oxide layer on carbon coated ANPs. Our results show that during the thermal oxidation of carbon-coated ANPs with ramping (300 - 1200 K), the oxidation rate can be delayed at a low temperature as compared to bare ANPs, whereas the oxidation rate can be accelerated at elevated temperatures ($T > 700$ K) because the carbon coating locally peels off at the outer surface of the ANPs. These results are consistent with experimental literature and provide insights into the role of the carbon coating during the ANP oxidation.

The Effect of Halogenation Group on Amyloid-Beta 40 Monomer Aggregation in Alzheimer's Disease Using Molecular Modeling

Computational Nanobio Technology (CNBT) Laboratory | School of Material Science and Engineering
Hanbyeol Jin[a], Joy Kim[a], Lauren Milner[a], Jin Eun Shin[a], Chong In Shin[a]| Dr. Jee, Sang Eun[a], Dr. Kwon, Inchan[b] Dr. Jang, Seung Soon [a]
a. School of Materials Science and Engineering, Georgia institute of Technology, 771 Ferst Drive, Atlanta, GA 30332
b. School of Materials Science and Engineering, Gwangju Institute of Science and Technology, Gwangju, S. Korea

CNBT LAB
alzheimer

Background

Alzheimer's disease (AD) is one of the most common types of degenerative dementia. The formation of neurotoxic plaque composed of amyloid beta (Aβ) fibrils has been found in a relatively high portion of patients' brains. Investigation about the mechanism of the Aβ aggregation found that the initial Aβ monomer structure misfolds to form soluble and insoluble fibrils as the disease progresses.

Erythrosine B (ER) is a well-known food dye approved by the FDA. Our collaborator, Dr. Kwon observed that ER inhibited the formation of Aβ fibril through *in vitro* experiment. He also found that the halogenation of the ER structure improved ER's ability to inhibit the formation of fibrils. However, the mechanism of halogenated ER has not been discussed yet.

Motivation & Purpose

By performing MD simulation, the mechanism of the structural change of the halogenated ER on the Aβ protein monomer could be explained. We investigated the efficiency of the modified versions of the candidate drug.

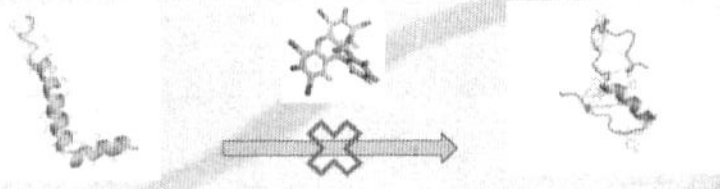

Initial Aβ 40 APS Aβ 40

Model Preparation

- Aβ40 from NMR studies in aqueous micellular environment (rscb.org, PDB ID: 1BA4)
- Determined binding site using AutoDock
- Modified ER: Eosin Y (EOY) modeled by Cerius2

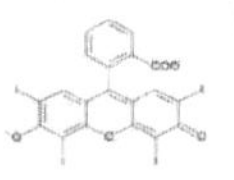

ERB

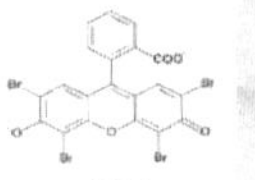

EOY

Method

- Simulation Set-up for MD

Software	Force Field	Water	Thermostat	Electrostatic
GROMACS (4.6.6)	AMBER99SB-ILDN	SPC	Nosé-Hoover Parrinello-Rahman	PME

- Equilibration Process
  - System neutralized with NaCl (150mM)
  - System equilibrated by NVT and NPT equilibration for 800ps each
  - Aβ(1-40), Aβ with ER, and Aβ with EOY in explicit water NPT 150.0 ns at T = 310K, P = 1 bar
- Simulation Set-up for DFT Calculation

Software	Functional theory	Basis	Charge Analysis	Total Charge	Spin multiplicity
Jaguar	B3LYP	6-31G**	Mulliken	-2	1

Result I. Structural Deformation of the Aβ40

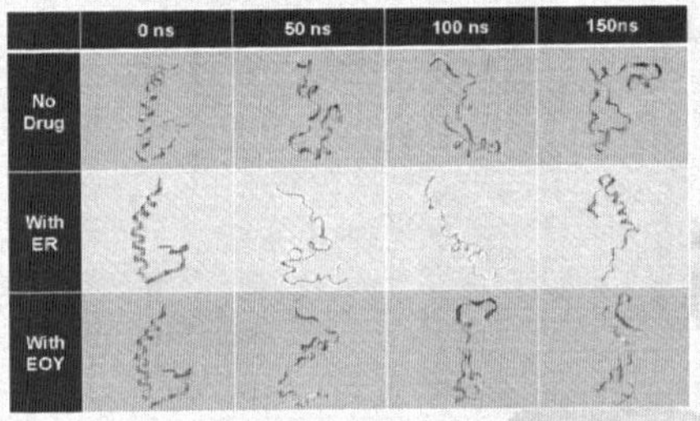

All three trajectories begin with an alpha helical structure, which is deformed into a hairpin loop when there are no drugs. The presence of ER prevent this deformation, which is known as the preceding states of the aggregation. However, with EOY, hairpin loop formation was observed at >100ns.

Result II. Statistical Analysis of Conformational Changes

Root-Mean-Square Deviation (RMSD) analysis was performed to see the change in atomic position of backbone of Aβ40 through time compared to the initial position.

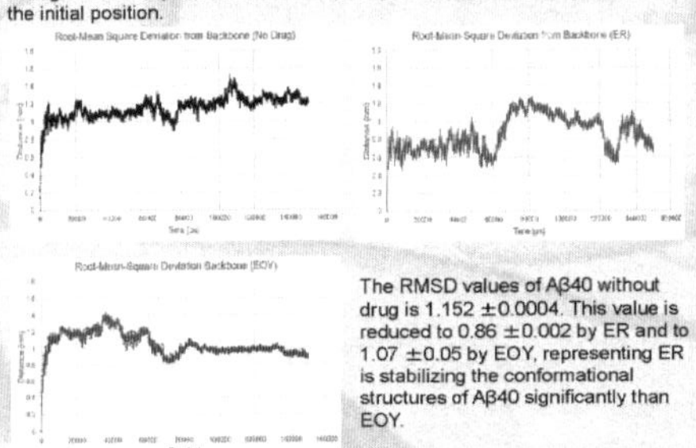

The RMSD values of Aβ40 without drug is 1.152 ±0.0004. This value is reduced to 0.86 ±0.002 by ER and to 1.07 ±0.05 by EOY, representing ER is stabilizing the conformational structures of Aβ40 significantly than EOY.

Result III. Contact Map between residues of Aβ40

Distance between Cα of each residue is plotted.

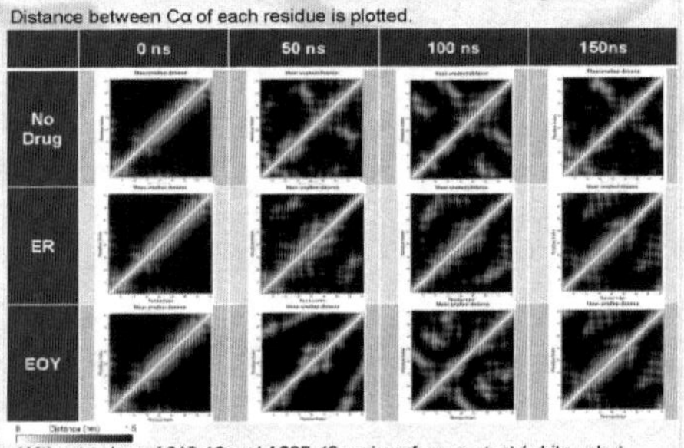

Without a drug, Aβ13-18 and Aβ35-40 regions form contact (white color) consistently over the simulation time, which corresponds to the hairpin loop formation in Result I. With ER, the contact map does not show a consistent pattern over time, meaning ER alters the formation of hairpin loop. With EOY, contact between Aβ2-4 and Aβ19-22 regions was observed at 100~150ns.

Result IV. Change in Secondary Structure of Aβ40 : DSSP analysis

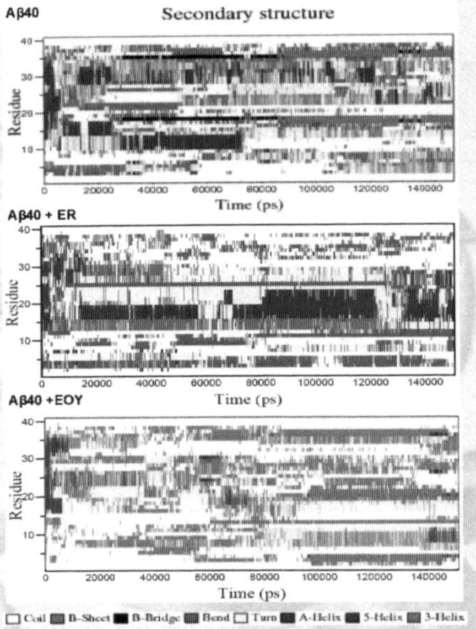

Without a drug, the alpha helical structure disappears and beta sheet structures form specifically at Aβ15-18 and Aβ35-38 regions, where the hairpin loop is formed.
When treated with ER, the alpha helical structure was maintained and beta sheet structure was not formed.
When treated with EOY, beta bridge formation between Aβ15-18 and Aβ35-38 regions is altered but beta sheet is formed at Aβ2-4 and Aβ19-22 regions, representing hairpin loop formation.

Result V. Distance between Drug and Aβ40

The distance between the drug and Aβ40 was calculated.

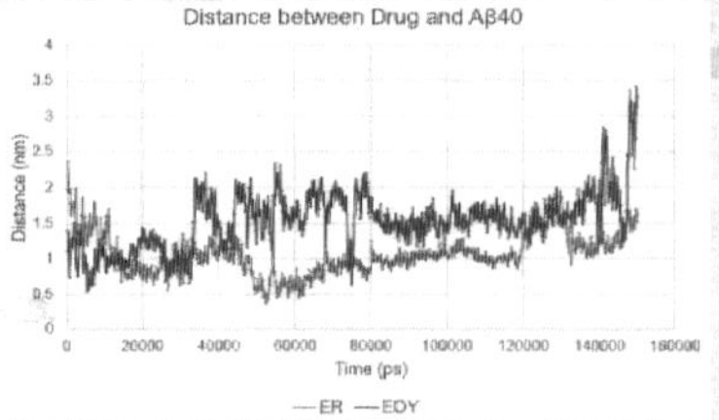

The distance between ER and Aβ40 is fluctuating around 1 nm over the simulation time, indicating binding of ER to Aβ40 is stable. The distance between Aβ40 and the EOY was significantly larger than that of ER and keeps increasing over time, which indicates that the binding of EOY to Aβ40 is unstable.

Summary & Conclusion

The purpose of this research is to study the effect of ER and its halogenation on Aβ40 monomer deformation.

Structural changes of Aβ40

Without drugs, Aβ40 is deformed to form a hairpin loop, which is known as the important factor of oligomerization, leading to the neurotoxicity. ER was effective in preventing hairpin loop formation but EOY could not prevent hairpin loop formation although the beta bridge formation was altered.

Statistical analysis of conformational changes

RMSD analysis was conducted to investigate the alteration of the structures over simulation time. The results showed that ER effectively reduced the RMSD of Aβ40 by stabilizing the structures, while EOY still alters the structures until it stabilizes at 1.0nm which refers to the structural deformation of Aβ40 monomer.

Contact map analysis

ER efficiently prevented the formation of the contact between Aβ13-18 and Aβ35-38 regions, which correspond to the hairpin loop formation. However, the contact map in presence of EOY showed hairpin loop formation in Aβ2-4 and Aβ19-22 regions.

Secondary structural change of Aβ40 monomer

DSSP (Definition of Secondary Structure of Proteins) Analysis

Without a drug, beta sheet structures are formed between Aβ15-18 and Aβ35-38 regions, which is part of the hairpin loop. When ER and EOY were added, they were able to allow the Aβ40 monomer to prevent the formation of the loop at these regions. However, with EOY, structures is altered significantly into the coil structures and forms beta sheet structures at Aβ2-4 and Aβ19-22 regions.

Minimum Distance Between Drug and Aβ40 Monomer

The distance from the center of mass of Aβ40 to the EOY is greater than that of the ER and keeps increasing over simulation time, which infers weaker binding energy. Due to this unstable binding, EOY was not efficient in preventing the beta sheet formation.

Conclusion

ER was effective in preventing the deformation of Aβ40. The key factors to prevent beta sheet formation of the Aβ40 were that the bound drugs prevented the formation of hairpin loops between Aβ15-18 and Aβ35-38 regions. However, with EOY, deformation of Aβ40 is observed which is assumed as halogenation of drug significantly alters binding affinity to Aβ40.

Future Study

Cerius²
AutoDock

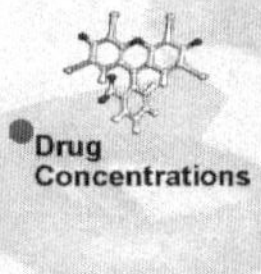

Future Analysis of Protein-Drug Docking

In the future, the effect of the halogenation on protein drug docking will be investigated. Investigation of the Docking of ER and EOY will be analyzed by the distance and interaction energy with Aβ40. Further, binding energy of drugs will be estimated using SMD(Steered Molecular Dynamics) simulations to elucidate the effect of the halogenation on drug- Aβ40 binding. Based on the observations, we will predict the optimal structural properties as efficient drugs to prevent the aggregation of Aβ40.

MD Simulation of Aβ40 and Norepinephrine Metabolites

We will collaborate with Dr. Jin Ryoun Kim's lab at NYU to understand how norepinephrine metabolites such as NMN and DHMA affect the aggregation and fibril formation of the Aβ 40 monomer. As with modified ER drugs, varied concentrations of metabolites will also be studied to understand their effects.

Contact Information

CNBT Website
http://cnbt.mse.gatech.edu

Hanbyeol (Clara) Jin
Undergraduate Student
Georgia institute of Technology
Chemical and Biomolecular Engineering
Email: hjin40@gatech.edu

Dr. Seung Soon Jang (Advisor)
Associate Professor
Georgia institute of Technology
Materials Science and Engineering
Email: SeungSoon.Jang@mse.gatech.edu

420076 Exploring of the Pre-Polymerization Coordination of 1-Vinylimidazole

Abstract:

In bulk photopolymerization, it can be difficult to maintain a high reaction rate and achieve complete conversion. To alleviate these challenges, several methods have been explored to improve the mass transfer of the monomers in the reaction mixtures. One of the most prevalent strategies is to increase the local monomer concentration in the mixture, both before polymerization and during the polymerization process. Since the thermochemical properties of the solution will change during polymerization, the availability of monomers to react with active sites, both before and during the reaction, will affect the overall reaction rate and conversion of the polymerization [1].

Poly(vinylimidazole) (poly-VIm) has a multiple range of applications, including: coating electrodes [2], adhesive development [3], fuel cell membranes [4], CO_2[5][6] and metal ion adsorption [7], and biomedical application[8]. Also, it is one of the ionic liquid (IL) coordinated polymers that has received a great deal of attention in radical polymerization. Previous experimental studies have indicated that cations (Zn^{2+}, AL^{3+}, Li^+, and Mg^{2+}) can increase both the reaction rate and the monomer conversion in radical polymerization, since the cations can form favorable complexes with the monomers [1]. While most of the past polymerization work has dealt with finding conventional solvents and salts to enhance polymerization [9], more recent research has considered ILs as promising additives for these reactions [10]. These recent experimental results indicate that polymerizable monomers coordinated with ILs, formed from weakly coordinating anions and cations, can enhance polymerization reaction kinetics [1]. Lithium bistriflimide ($LiTf_2N$) is one of the ILs that has been investigated experimentally in the polymerization of VIm. This IL had a substantial effect on increasing reaction kinetics, since mass transfer is a major limitation in this photopolymerization reaction. The addition of $LiTF_2N$ leads to 100% monomer conversion in the 1VIm: $1Tf_2N$ solution, as reported by Whitley et al. [1]. Different concentrations of $LiTf_2N$ with VIm show interesting behavior. Though the initial reaction rate of the 1:1 (VIm:$LiTf_2N$) solution is lower than neat VIm polymerization, the final monomer conversion for the 1:1 solution is higher and reaches 100% conversion after about 30 min. The initial reaction rate for the 1:1 molar concentration is lower in comparison to the other molar concentrations (2:1 and 3:1) [1]. This observed behavior (higher conversion and lower initiating reaction) needs to be fundamentally explored. Currently, there is no direct explanation for this phenomenon, and the molecular-level interactions in this type of system are unknown. While Li^+ has been experimentally found [1] to interact with 3 Tf_2N^- anions, we have conducted a thorough investigation of the Li^+ cation interactions in the VIm:$LiTf_2N$ system, in comparison to another cation (Na^+) and other anions (BF_4^- and PF_6^-). In order to develop a broad understanding of the polymerization dynamics, polymer structure, and properties, the molecular-level pre-polymerization structure of this mixture needs to be clearly quantified. In this work, molecular dynamics simulations of different VIm:$LiTf_2N$ solutions show that Li^+ locally enriches key reaction sites involved in polymerization. This local enhancement of site-site interactions is suggested to play a vital role in the experimentally observed improvement in polymerization behavior. References:

[1] Whitley J.W, Horne W.J, Danielsen S.P.O.; Shannon M.S, Marshall J.E, Hayward S.H, Gaddis C.J, Bara J.E. Enhanced photopolymerization rate & conversion of 1-vinylimidazole in the presence of lithium bistriflimide. *European Polymer Journal* **2014,** *60*, 92-97.

[2] Yildiz G, Oztekin N, Orbay A, Senkal F. Voltammetric determination of nitrite in meat products using polyvinylimidazole modified carbon paste electrode. *Food Chemistry* **2014,** *152*, 245-250.

[3] Fink JK. Handbook of engineering and specialty thermoplastics: water soluble polymers. Salem, MA: Scrivener: 2011.

[4] MacAodha D, Conghaile P.O, Egan B, Kavanagh P, Leech D. Membraneless glucose/oxygen enzymatic fuel cells using redox hydrogel films containing carbon nanotubes. *ChemPhysChem* **2013,** *14* (10), 2302-2307.

[5] Farjaminezhad M, Tehrani M.S, Azar P.A, Hussain S.W, Bohlooli S. Polyvinylimidazole/sol-gel composite as a novel solid-phase microextraction coating for the determination of halogenated benzenes from aqueous solutions. *Journal of Separation Science* **2014,** *37* (12), 1475-1481.

[6] Corazza M.Z, Ribeiro E.S, Segatelli M.G, Tarley C.R.T. Study of cross-linked poly(methacrylic acid) and polyvinylimidazole as selective adsorbents for on-line preconcentration and redox speciation of chromium with flame atomic absorption spectrometry determination. *Microchemical Journal* **2014,** *117*, 18-26.

[7] Wang R.X, Men J.Y, Gao B.J. The adsorption behavior of functional particles modified by polyvinylimidazole for Cu(II) ion. *Clean-Soil Air Water* **2012,** *40* (3), 278-284.

[8] Anderson EB, Long TE, Imidazole- and imidazolium-containing polymers for biology and material science applications. *Polymer* **2010,** *51* (12), 2447-2454.

[9] Gromov V.F, Osmanov T.O, Khomikovskii P.M, Abkin A.D. Polymerization of acrylamide in various solvents in the presence of lewis acids. *European Polymer Journal* **1980,** *16* (9), 803-808.

[10] Pedron S, Guzman J, Garcia N. Polymerization kinetics of ethylene oxide methacrylates in ionic media. *Macromolecular Chemistry and Physics* **2011,** *212* (8), 860-869.

420164 Local Hydration Structures and Dynamic Properties of Tetra-Alkyl Ammonium Aqueous Solution Systems: An Insight from Molecular Dynamics Simulations

Abstract:

Understanding the behavior of aqueous solutions containing tetra-alkyl ammonium cations is of great importance in a number of applications, including polymer membranes for fuel cells. We have developed and validated new polarizable force field for atomistic molecular dynamics (MD) simulations of aqueous solutions containing tetra-methyl ammonium (TMA) and tetra-butyl ammonium (TBA) cations with OH^- or Br^- anions. The force field was parameterized against density functional theory calculations (M052X/aug-cc-pvDz) of binding energies and optimal geometries of different molecular clusters. Extensive MD simulations of TMA/Br/water, TBA/Br/water and TMA/OH/water systems were conducted as function of solution composition (ion pairs-water molar ratios of 1:10, 1:20, 1:30, 1:63, 1:500). The comparison of ions local coordination shell and analysis of species self-diffusion coefficients showed a very good agreement with existing experimental data for Br anion-containing systems, especially at low ion concentrations, where previous simulations failed to reproduce consistent trends with experiments. The differences in self-diffusion coefficients of species between the experimental data and our simulations results are within 10%-15%.

422187 Insight into Selective Catalytic Reduction on Cu-SSZ-13 from Molecular Simulation

Abstract:

Insight into Selective Catalytic Reduction on Cu-SSZ-13 from Molecular Simulation

Christopher Paolucci[1], Atish A. Parekh[2], Ishant Khurana[2], John Di Iorio[2], Hui Li[1], Trunojoyo Anggara[1], Jeffrey T. Miller[2], Rajamani Gounder[2], W. Nicholas Delgass[2], Fabio H. Ribeiro[2], and William F. Schneider[1]*

[1]Department of Chemical and Biomolecular Engineering, University of Notre Dame, Notre Dame IN 46556 (USA)

[2]School of Chemical Engineering, Purdue University, West Lafayette, IN 47907 (USA)

The selective catalytic reduction (SCR) of NO_x in excess oxygen is a key challenge in meeting increasingly stringent emission regulations.

$$4\,NH_3 + 4\,NO + O_2 \rightarrow 4\,N_2 + 6\,H_2O \quad (SCR)$$

The metal-exchanged zeolite Cu-SSZ-13 in particular exhibits sufficient activity and hydrothermal stability for practical applications. **A primary obstacle for further improvement of SCR catalysts is a lack of molecular-level understanding of the active sites and mechanism**. Here, using Monte Carlo (MC) simulation, density functional theory (DFT) calculations, *ab-initio* molecular dynamics (AIMD), and *ab-initio* thermodynamic modeling we show that:

- SCR occurs on isolated metal sites in zeolites. The types and maximum number of these sites are estimated through a MC/DFT method[1]. This result is experimentally validated through a variety of techniques including temperature programmed desorption (TPD), X-ray absorption spectroscopy (XAS), UV-Visible spectroscopy, and kinetic experiments on zeolites of varying compositions.[2]

- Develop a redox mechanism that describes the SCR reaction at a molecular level on these sites based off of AIMD simulations and DFT calculations. The mechanism and its intermediate species are validated through XAS experiments at Argonne national laboratory and acid site titration experiments at Purdue University.[3]

- Model the kinetics of SCR and validate the model against *operando* XAS experiments at a variety of catalytic conditions.

We show SCR occurs through a redox mechanism on all isolated Cu sites and identify the key reaction steps in the reduction and oxidation half cycles [3].

References

[1] Bates SA, Verma AA, Paolucci C, Parekh AA, Anggara T, Yezerets A, Schneider WF, Miller JT, Delgass WN, Ribeiro FH. Identification of the active Cu site in standard selective catalytic reduction with ammonia on Cu-SSZ-13. Journal of Catalysis. 2014;312:87–97.

[2] Verma AA, Bates SA, Anggara T, Paolucci C, Parekh AA, Kamasamudram K, Yezerets A, Miller JT, Delgass WN, Schneider WF, Ribeiro FH. NO oxidation: A probe reaction on Cu-SSZ-13. Journal of Catalysis. 2014;312:179–190.

[3] Paolucci C, Verma AA, Bates SA, Kispersky VF, Miller JT, Gounder R, Delgass WN, Ribeiro FH, Schneider WF. Isolation of the Copper Redox Steps in the Standard Selective Catalytic Reduction on Cu-SSZ-13. Angewandte Chemie International Edition. 2014;53(44):11828–11833.

420519 Assessment and Improvement of Methods for Computing Net Atomic Charges in Periodic and Nonperiodic Materials

Abstract:

Net atomic charges are widely used to concisely convey information about the distribution of electrons among atoms in materials. Here, we compare the performance of various methods for computing net atomic charges in periodic and non-periodic materials. In all, we examined a diverse set of more than a hundred different materials, including molecules, ions, solid surfaces, porous solids, non-porous covalent and ionic solids, organometallic complexes, nanoclusters, nanotubes, monolayer sheets, polymers, and large biomolecules (e.g., DNA). We introduce a new and improved charge assignment method (called DDEC5) whose net atomic charges: (a) have excellent conformational transformability, (b) follow electronegativity trends, (c) are within a chemically acceptable range (e.g., net atomic charge less than or equal to +1 for Na atoms in materials), (d) approximately reproduce the electrostatic potential surrounding the material, and (e) often correlate to spectroscopically measured core-electron binding energies. These properties make DDEC5 net atomic charges ideally suited for constructing flexible force-fields and for studying the properties of complex materials. In addition, we use this method to compute atomic spin moments in collinear and non-collinear single molecule magnets.

423178 Molecular Dynamics Simulations of Mixtures of Refrigerants and Deep Eutectic Solvents

Abstract:

Molecular dynamics simulations of mixtures of refrigerants and deep eutectic solvents

Rubaiyet Abedin, John C. Flake and Francisco R. Hung

Cain Department of Chemical Engineering, Louisiana State University, Baton Rouge, LA 70803

Heating and cooling buildings in the U.S. consumes an enormous amount of energy (>10 quadrillion BTU), and is responsible for adding ~1 billion metric tons of CO_2 in the earth's atmosphere every year. Much of this energy is used as electricity in vapor-compression systems; however, this technology is mature and only evolutionary improvements are expected in the near future. Remarkably, a few studies have shown that several common ionic liquids (ILs) can be combined with standard fluorocarbon refrigerants for use in absorption refrigeration systems that use waste heat at relatively low temperatures (~ 100 °C). Nevertheless, there is limited understanding (and data) on the VLE behavior of these systems, and only one working example of an absorption system using this type of mixture. Furthermore, deep eutectic solvents (DESs), a relatively new class of solvents, share many of the properties of ILs while being considerable cheaper and mostly nontoxic. A fundamental understanding of how the chemical structure of the different species affects the solubility of fluorocarbons in a DES is crucial to design mixtures suitable for use in absorption refrigeration systems that use solar energy or waste heat. In this study, we used molecular dynamics (MD) simulations to study mixtures of a conventional fluorocarbon refrigerant, 1,1,1,2-tetrafluoroethane (R134a) with two conventional DESs (1:2 choline chloride/urea and 1:2 choline chloride/glycerol). Results for the Henry's law constant of R134a in the two DESs as a function of temperature are reported and analyzed in terms of molecular level properties (radial distribution functions, relevant interaction energies).

422757 Molecular Modeling of Homogeneous Nucleation of [Bmim+][Cl-] from the Melt

Abstract:

The study of solid phases of ionic liquids (ILs) confined inside nanoporous materials is relevant for the development of IL-based nanomaterials with unique properties (fluorescence, magnetic susceptibility). Although nucleation is a critical step in the process of crystallization, a fundamental understanding of these phenomena at the molecular level is still challenging, especially in the field of ILs. 1-butyl-3-methylimidazolium chloride, $[bmim^+][Cl^-]$, a prototype ionic liquid, has a relatively simple structure and can form two crystal polymorphs. Here we performed molecular simulation to study the homogeneous nucleation of $[bmim^+][Cl^-]$ from its supercooled liquid phase to its orthorhombic or monoclinic crystalline structure. The string method in collective variables[1,2] was used in combination with Markovian milestoning with Voronoi tessellations[2,3] and order parameters for molecular crystals[4] to sketch a minimum free energy path connecting the supercooled liquid and the crystal phases, and to determine the rates involved in the nucleation process. The mechanism of nucleation and the free energy barriers associated with this process will be presented and discussed.

[1] L. Maragliano, A. Fischer, E. Vanden-Eijnden and G. Ciccotti, "String Method in Collective Variables: Minimum Free Energy Paths and Isocommittor Surfaces", *J. Chem. Phys.* **2006**, *125*, 024106

[2] V. Ovchinnikov, M. Karplus and E. Vanden-Eijnden, "Free Energy of Conformational Transition Paths in Biomolecules: The String Method and Its Application to Myosin VI" *J. Chem. Phys.* **2011**, *134*, 085103

[3] E. Vanden-Eijnden and M. Venturoli, "Markovian Milestoning with Voronoi Tessellations", *J. Chem. Phys.* **2009**, *130*, 194101

[4] E. E. Santiso and B. L. Trout, "A General Set of Order Parameters for Molecular Crystals", *J. Chem. Phys.* **2011**, *134*, 064109

423417 Equilibrium Properties of DNA Confined in Nanochannels: A Monte Carlo Chain Growth Approach

Abstract:

Recent developments in next-generation sequencing (NGS) techniques have opened the door for low-cost, high-throughput sequencing of genomes. However, these developments have also exposed the inability of NGS to track large-scale genomic information, which is extremely important to understand the relationship between genotype and phenotype. Genome mapping offers a reliable way to obtain information about large-scale structural variations in a given genome. A promising variant of genome mapping involves confining single DNA molecules in nanochannels whose cross-sectional dimensions are approximately 50 nm. Despite the development and commercialization of nanochannel-based genome mapping technology, the polymer physics of DNA in confinement remains poorly understood. In this poster, we show results of our studies on the equilibirum and near-equlibrium properties of DNA confined in nanochannels. We performed large-scale Monte Carlo simulations of a coarse-grained model of DNA via the pruned-enriched Rosenbluth method (PERM). In addition, we combined our PERM simulations with computational fluid dynamics (CFD) calculations of the hydrodynamic tensor to estimate the diffusivity of confined DNA. We elucidate our results in the context of existing theories in polymer physics, and discuss how our work can be used to engineer better conditions for genome mapping technology.

424820 The Mechanical Property of Dental Resin Composite Consisting of Bisgma and Tegdma Using Molecular Dynamics Simulations

Abstract:

Abstract

Improvements in dental composite materials, particularly in resin-based filling composite, have modernized the field of dentistry. Dental resin composites, which are tooth-colored materials used as adhesive or restorative materials, have replaced a mercury in dental amalgam filling due to aesthetic appeal. Most dental resin composites are composed of bisphenol-A-glycidyl-dimethacrylate (BisGMA) and triethyleneglycol-dimethacrylate (TEGDMA). Although these are beneficial advances, there are still limitations on the use of composites in dental restoration. Accordingly, our research team have put an increasing effort on researching the mechanical properties of dental composites resins.

Monomer complex composed of 50:50 BisGMA and TEGDMA was created using Cerius2 and geometry optimization was performed by Jaguar. Molecular dynamic simulation of compression and extension was performed with strain of 50 %. To investigate the effect of timescale, the simulation was performed for 1 nanosecond and 2 nanoseconds.

As a result of our simulation, significant differences in mechanical properties, such as density, pressure, stress, and radial distribution function (RDF), were observed due to compression and extension. Furthermore, it was demonstrated that TEGDMA was a more flexible structure than BisGMA which contributed more to the deformation of the original resin structure. For our team's future plan, torsion angle change will be analyzed to further investigate the deformation of the original structure.

425264 Transport of Protein Shaped Colloids through Nanopores of Different Shapes

Abstract:

The development of devices employing protein transport through complex geometries is a critical problem in the fields of molecular separation, detection, analysis, and drug delivery. In order to observe dynamic behavior of such systems on relevant timescales using simulations, a newly developed coarse graining model for proteins has been employed to study transport in pores. In this model, a rigid surface of coarse-grain beads are spread evenly over the solvent accessible surface of a protein and assigned charges based on potential mapping of atomistic models. The hydrodynamics are resolved using lattice Boltzmann fluid. The computation effort is therefore vastly reduced while maintaining accurate descriptions of the excluded volume, electrostatic, and hydrodynamic effects that dominate protein transport. The model has been used to observe protein transport under confinement at various pore geometries and at different pH and ionic strength values.

425754 Molecular Simulation of the Partitioning Phenomena of Oil and Dispersant Components in Air-Seawater-Oil System

Abstract:

As majority of research on chemical dispersants, their interactions with oil, and the effects that dispersant-oil mixtures has been focused primarily on toxicology, yet studies on molecular scale phenomena are important to obtain a fundamental understanding of how the components of chemical dispersants and oil interact at different interfaces, such as the air-seawater and the oil-seawater. Here we report simulation results of the partitioning of several Corexit components [i.e. sodium dioctyl sulfosuccinate (DOSS), Span 80, as well as solvents such as dipropyl glycol butyl ether (DPGE)] between model seawater and oil phases especially heavy oil. These results are relevant to biodegradation and toxicological studies. Experimental researchers simulate the mixing of fresh oil, seawater, and dispersant using a prescribed lab protocol; the resulting aqueous phase (termed chemically-enhanced water accommodated fraction, or CEWAF) is then separated from the oil phase and used for biodegradation or toxicology exposure studies in the lab. However, preliminary results suggest that CEWAFs prepared using the prescribed lab protocol are representative of only a small subset of environmental conditions, which do not include oil spills in the Arctic (e.g., Exxon Valdez), oil weathering effects, or the direct injection of Corexit in the turbulence of the Macondo wellhead in the Gulf of Mexico. Therefore, as part of our initial efforts in this topic, molecular dynamics (MD) simulation is performed to study the partitioning of Corexit components between model seawater and oil phases. Results and discussion will be presented.

425764 Reaxff Reactive Molecular Dynamics Simulations with Explicit Electrons and Applications to Battery Interfaces

Abstract:

Interfacial chemistry at the electrode-electrolyte interfaces are crucial for Li-based rechargeable battery performance. In typical Lithium ion batteries (LIBs) with the most prevalent carbonate type electrolytes, anode-electrolyte interface undergoes reduction reaction, while oxidation occurs at the cathode-electrolyte interface. The formation of the cathode/electrolyte and anode/electrolyte interface and the stability and diffusion properties of this interface layer are of crucial importance for life-cycle, power density, and safety of LiBs. This interfacial chemistry involves explicit electron flow and redox reactions. Any theoretical method to elucidate this intricate chemistry requires the capability of describing electron flow, and associated redox reactions. However, at this moment, no single computational method contains the capability to fully simulate the electron flow/chemistry coupling. Quantum mechanics based methods (QM) have the ability to simulate explicit electrons, but the requirement of extremely high length and time scale limits their usage to couple the complex chemistry at the cathode/anode interfaces to the electron flow. On the other hand, parameterization of the most of the force field methods are aimed to coarse grained, therefore, access to the electronic structures are usually absent. In the genre of reactive force fields, current version of ReaxFF method also possesses limitations in studying redox reactions. Therefore, an effective way for investigating interfacial chemistry is to extend the capability of the ReaxFF method to incorporate explicit electron- or hole-description for studying reduction and oxidation reactions, respectively. The extension of the ReaxFF method towards explicit electron- or hole-particle concept requires introducing new energy terms as well as redefinition of some of the current energy functional. In this study, we employed this newly developed electron version of ReaxFF (eReaxFF) for training our force field to capture electron affinities and ionization potential of various species. We are studying electron dynamics on different molecules and validating our findings against Ehrenfest dynamics results. We also apply this method to describe ionic liquid systems. The basic concepts of explicit electron treatment in eReaxFF, results from force field fitting and MD simulations will be presented.

425766 Developing Reaxff Force Field to Study Syngas Combustion Kinetics and Extending It for Larger Hydrocarbons

Abstract:

Syngas (Synthesis Gas); a mixture of CO and H_2gas, has shown a great potential to solve future energy crisis. However, a detail chemical kinetic study on syngas combustion is required to ensure quality combustion and identify the key reactions. Although numerous kinetic models are proposed, many of the reaction pathways and their relative importance are yet to be known. To address this issue, we developed a ReaxFF interatomic potential to describe the C/H/O system particularly for smaller hydrocarbons, as the previous ReaxFF combustion force field [1]while successful for large hydrocarbon pyrolysis and combustion, was less accurate for C1-chemistry.

Molecular Dynamics (MD) simulations performed on a syngas system using this force field can capture many more reaction channels than previous force field, specially the reactions involving HCO radicals. HCO is believed to be the most important intermediate in syngas combustion and adding the reactions involving HCO with H_2-O_2system kinetics can completely define the kinetic model for syngas. Previous ReaxFF combustion force field estimated a high-energy barrier for hydrogen atom and CO combination reaction, leaving this specie and the associated reaction channels absent in the dynamics-in contrast to DFT-results that predict a low barrier for this reaction. Our current force field resolves this issue and our result demonstrate the consumption of HCO radicals through a couple of well-defined chain terminating reactions. Overall, the simulation results can describe the syngas combustion kinetics that is in good agreement with literature.

Once the force field to capture the C_1chemistry in combustion kinetics is completely validated, it will be integrated with the existing ReaxFF combustion force field for larger hydrocarbons [2]–[5], which has been immensely successful in recent years. This will result a fully transferable ReaxFF C/H/O description, applicable for small to large-scale combustion of wide variety of hydrocarbons.

-

Reference:

[1] **Chenoweth K.,** ***et al.*** , "Development and Application of a ReaxFF Reactive Force Field for Oxidative Dehydrogenation on Vanadium Oxide Catalysts," *J. Phys. Chem. C*, **112**(37), pp. 14645–14654 (Sep. 2008).

[2] **Castro-Marcano, F.,** ***et al.***, "Combustion of an Illinois No. 6 coal char simulated using an atomistic char representation and the ReaxFF reactive force field," *Combust. Flame*, **159**(3), pp. 1272–1285 (Mar. 2012).

[3] **Qian, H.-J.,** ***et al.***, "Reactive Molecular Dynamics Simulation of Fullerene Combustion Synthesis: ReaxFF vs DFTB Potentials," *J. Chem. Theory Comput.*, **7**(7,) pp. 2040–2048 (Jul. 2011).

[4] **Liu, L.,** ***et al.***, "Mechanism and Kinetics for the Initial Steps of Pyrolysis and Combustion of 1,6-Dicyclopropane-2,4-hexyne from ReaxFF Reactive Dynamics," *J. Phys. Chem. A*, **115**(19), pp. 4941–4950 (May 2011).

[5] **Chenoweth, K.,** ***et al.***, "Initiation Mechanisms and Kinetics of Pyrolysis and Combustion of JP-10 Hydrocarbon Jet Fuel," *J. Phys. Chem. A*, **113** (9), pp. 1740–1746 (Mar. 2009).

426539 Molecular Dynamics Simulation of Lipid Bilalyer Consisting of DPPC and Mppc: Effect of Configuration

Abstract:

Liposomes, spherical nanoparticles composed of phospholipid bilayers, have been suggested as a drug delivery system to transport chemotherapeutics. Through the transition of lipid bilayer from gel to liquid-crystalline, the structure reaches its maximum permeability and thereby the encapsulated drugs are released. Incorporating both Dipalmitoyl Phosphatidyl Choline (DPPC) and a mono-tailed lysolipid (MonoPalmitoyl Phoshatidyl Choline, MPPC)—the transition temperature could be targeted at the temperature of mild hyperthermia. There have been studies that looked at the different percent compositions of the mixed lipid bilayer systems. Yet, there has been a lack of understanding on how distribution of the lipids affects the overall structure of the mixed lipid system. Therefore, this study runs fully atomistic molecular dynamics (MD) simulations of the lipid bilayer systems with two different configurations: dispersed and segregated model. Dispersed model has lysolipid molecules evenly distributed across the structure whereas segregated model has lysolipid molecules aggregated in the middle of the structure. The system is comprised of 10 percent MPPC and 90 percent DPPC. The changes in structure through the phase transition are investigated.

427167 Enhanced Oxygen Incorporation Near the Grain Boundary on Yttria-Stabilized Zirconia: A Density Functional Theory Study

Abstract:

Solid oxide fuel cell (SOFC) is a clean technology with less emission of air pollutants, but the high operating temperature (600~1000°C) causes significant issues related to materials stability and system cost. Since electrolyte conductivity and electrode kinetics are considerably deteriorated at lower temperature, improving the performance on them plays an important role in reducing the operating temperature of SOFC. The recently reported oxygen incorporation enhancement near the grain boundary (GB) of yttria-stabilized zirconia (YSZ) accelerated the realization of low temperature SOFC. However, the detailed reaction mechanism of the process is still not clear yet. To unravel it, we performed first-principles calculations to quantitatively access each elementary step of oxygen incorporation into YSZ, including yttrium dopant segregation, oxygen vacancy formation, and oxygen adsorption. We found that the doped yttrium which preferentially segregates towards GB facilitates to create the formation of oxygen vacancy between the yttrium pair. The oxygen is then easily adsorbed near the oxygen vacancies accumulated at GB, eventually incorporating into the vacancy sites. Our results will provide useful insight to unravel oxygen incorporation mechanism at the interfaces on defected oxide materials.

429668 Combining Spectroscopy Experiments and Molecular Simulation to Determine Structural and Mechanistic Details of Adsorbed Biomolecules

Abstract:

Combining spectroscopy experiments and molecular simulation to determine structural and mechanistic details of adsorbed biomolecules

Kayla Sprenger[1], Tobias Weidner[2], Jim Pfaendtner[1]

[1]*Department of Chemical Engineering, University of Washington, Seattle, WA, 98195, USA*

[2]*Max Planck Institute for Polymer Research, Mainz, 55128, Germany*

ABSTRACT

Proteins and the mechanisms they mediate at solid/liquid interfaces play a critical role in many processes including surface fouling, biocatalysis, and biomineralization. Despite their importance, we often lack a molecularly detailed understanding of interfacial mechanisms such as protein recognition, binding, and surface-induced conformational change. While atomistic simulations of protein adsorption have strong potential to combine with experiments to elucidate structural and mechanistic details at the interface, challenges of timescale limitations and strong protein/surface binding have prohibited the progress of such simulations. To-date there are virtually zero structures of a biomolecule adsorbed to a surface that have been made accessible in the Protein Data Bank. For the first time results of molecular simulation, using the enhanced sampling method PTMetaD-WTE[1] to overcome aforementioned challenges, have been combined with experimental data from solid-state nuclear magnetic resonance (ssNMR), near edge X-ray adsorption fine structure (NEXAFS), and sum frequency generation spectroscopy (SFG) to determine the structure and orientation of the SN15 binding domain of the salivary protein statherin on hydroxyapatite (HAp). Additionally, multiple SN15 peptides have been simulated to represent more closely the experimental conditions of a crowded microenvironment at the interface. Resulting structural information is compared to the single protein adsorption results to determine the role of lateral protein-protein interactions. From this comparison, mechanistic details of statherin film formation and its control of HAp growth are hypothesized. This poster will also propose some general principles for efficiently simulating with metadynamics and computing experimental observables for protein adsorption systems.

References

[1] Deighan, M.; Bonomi, M.; Pfaendtner, J. Efficient Simulation of Explicitly Solvated Proteins in the Well-Tempered Ensemble. *JCTC* **2012**, *8*, 2189-2192.

426676 Density Functional Theory Study of CO2 Adsorption on Surface-Modified Graphene with Nitrogen Heterogeneity

Abstract:

Emission of carbon dioxide, during anthropogenic activities such as energy production from fossil fuels and usage of transportation, has been continuously increasing. Due to its detrimental effects to the ecosystem, researches for reducing carbon dioxide are being actively conducted. Among various capture technologies, adsorption is advantageous in terms of regenerability of adsorbents and high efficiency. Since capture takes more than 70% of the cost of carbon dioxide capture and storage (CCS) technology, highly efficient adsorbents are needed to be developed. In this respect, density functional theory (DFT) can be exploited to analyze adsorbents and understand the requirements for good adsorbents by confirming interaction of adsorbents with carbon dioxide in nanoscale. In this study, based on DFT, calculations for theoretical design of adsorbent have been conducted. Among many carbon-based adsorbents, graphene was chosen for study due to its high surface area and unique two dimensional structure. The effects of heteroatom dopant and functionalities, especially nitrogen, have been investigated. Geometry optimization followed by carbon dioxide binding energy calculation made it possible to compare various N functionalities and to confirm the positive effect of N doping. From our results, it is expected that the optimal conditions of the surface of adsorbents can be figured out to aid practical synthesis of adsorbents.

429888 Microscopic Diffusion of CO2 in Clay Nanopores

Abstract:

Natural gas production from shale in recent years has made a significant impact on the energy sector. It drastically changed the energy mix of the United States, making it much less reliant on coal, and hence helping its transition into cleaner energy sources. At the same, it also presents potential opportunities to mitigate global warming by using shale reservoirs for CO_2 sequestration or using $CO_2$for enhanced natural gas recovery.

Transport and fluids flow in conventional rocks is considered a mature subject, however, the shale's nonporous nature challenges the common understanding of fluids and rocks. Many of the commonly used models, such as Darcy law, are no longer applicable due to the heterogeneous pore structure, which ranges from micro to mesopores. Shale is composed of two main media, organic and inorganic. In the past decade, many studies have been dedicated to the investigation of the organic part of coals as part of enhanced coal bed methane recovery, which has some resemblance to the organic part of shale. These studies were used as a basis to further study and elucidate the role of the organic matter in shale. However, there is still distinction between coal and shale and much remains not understood. Less attention has been paid to the clay part of shales. Illite, the dominant clay material in shales, has not been the subject of many scientific articles.

Molecular simulation techniques have proven to be a powerful tool in in many fields, including materials science, biology, and physics. However, they have yet to be widely utilized in the oil and gas community. Physical understanding of shale at the molecular level is critical to maximizing the value of shale. This work focuses on using equilibrium molecular dynamics to study the impact of the microporous inorganic matter of shale on CO_2 diffusion. An illite model based on *ab initio*calculations is used to represent the clay matrix of shale. The canonical ensemble (NVT) was used to carry out simulations with ClayFF as the force field of choice. The Einstein relation based on a random-walk movement was used to determine diffusion coefficient. Simulations were run at a temperature range of 300-500K, which resembles subsurface conditions. Pore sizes in the 10-30A range were tested. Results show that confinement has a profound impact on diffusion. The diffusion coefficient was also found to be anisotropic due to the nature of the clay surface.

430562 Force Field Development of Chloroethenes

Abstract:

In this present effort, all atom force field are developed for chloroethenes from fitting the Lennard-Jones parameters to the pure component fluid properties. The all atom representation allows finding more accurate description of interaction potentials. Following the TraPPE force field protocol, we report Gibbs ensemble Monte Carlo simulations of the chloroethenes to find Lennard-Jones force field parameters by fitting the vapor liquid coexistence curves of the neat components. The simulated liquid densities and vapor pressures for chloroethenes show satisfactory agreement with the experimental data. In addition, simulations results for binary mixture of trichloroethylene and 1-propanol at different pressure are in excellent agreement with the experimental data.

430955 Mesostructured Self-Assembly of 12-Hydroxystearate in Organic Solvents

Abstract:

Derivatives of R-12-hydroxystearate are currently the main surfactants used in most high-performance grease applications. While hydroxystearates are known to form organogels, the detailed structure and formation mechanisms of those gels are not understood, leaving a gap in our understanding of the links between molecular structure and grease performance. This project addresses this gap by using united-atom molecular dynamics simulation to investigate the self-assembly of enantiomerically pure R-12-hydroxystearate in hexane. In three separate 1-microsecond-long MD simulations beginning from random configurations, we find that self-assembly is driven by two types of hydrogen bonding: the acetic-acid dimer structure between stearate carboxylates and 12-hydroxyl-to-12-hydroxyl H-bonds. The self-assembly proceeds through formation of mesostructures that are predominantly polarized five- and six-membered rings of molecules that assemble into lamellae that adhere to one another through a registry of acetic-acid ring-to-ring dimerizations. A simplified kinetic model is fit to the evolution of multimeric species observed directly in the simulations and provides rate constants that should one day prove useful in continuum-level modeling of elastohydrodynamic friction.

431530 Online Tools for the Trappe Family of Force Fields

Abstract:

The Transferable Potentials for Phase Equilibria (TraPPE) family of force fields is known for balancing accuracy and computational efficiency and for being applicable to a wide range of chemical compounds, state points, and thermophysical properties. With an increasing diversity of offerings (e.g., united-atom, explicit hydrogen, polarizable, and coarse-grained versions) and a growing user base, the TraPPE family of force fields has reached a level of complexity that benefits from a centralized method of access. Here we present the main features of the TraPPE website (http://www.chem.umn.edu/groups/siepmann/trappe/) [1] and highlight recent updates that allow full access to all published functional forms, parameters, simulated properties, and other resources that assist users in the successful implementation and continued use of these force fields. These updates include: search methods that allow users to find parameters and properties for specific compounds of interest, web-based tools for assembling new force field models (i.e., for compounds not simulated during the parameterization and verification process) using existing TraPPE-United Atom building blocks, access to new validation data for older TraPPE models, and the ability to request parameters for novel compounds.

[1] B.L. Eggimann, A.J. Sunnarborg, H.D. Stern, A.P. Bliss, and J.I. Siepmann, An online parameter and property database for the TraPPE force field, *Molec. Simul.*, **40**, 101-105 (2014).

430464 Rapid Computation of Thermodynamic Properties over a Large Multidimensional Space of Nonbonded Parameters

Abstract:

We show how thermodynamic properties can be computed over a large, multidimensional parameter space by combining multistate reweighting analysis with a linear basis function approach. This approach reduces the computational cost to estimate thermodynamic properties for over 130,000 tested parameter combinations from over a thousand CPU years to less than one CPU minute. The speed increase is achieved primarily by computing the potential energy as a linear combination of basis functions, computed as the difference of energy between two reference states, which can be done without any simulation code modification. We estimate thermodynamic properties with the Multistate Bennett Acceptance Ratio (MBAR) over the high dimensional space without the need to define *a priori* thermodynamic paths connecting the states. Instead, we adaptively sample a set of points in parameter space to create mutual phase space overlap. Regions of poor phase space overlap are detected by analyzing the eigenvalues of the sample states' overlap matrix. The phase space overlap to sampled states is monitored alongside the mean and maximum uncertainty to determine convergence as neither the uncertainty nor the phase space alone is a sufficient metric of convergence. The adaptive sampling scheme is demonstrated by finding with high precision the solvation free energies, entropies, and enthalpies of all charged particles of Lennard-Jones plus Coulomb functional form with charges between -2 and +2 and any physical values of ε_{ij} and σ_{ij} in TIP3P water.

431833 Atomistic Simulations of Unfolded and Intrinsically Disordered Proteins

Abstract:

Intrinsically disordered proteins (IDPs) do not fold to a single three-dimensional structure under physiological conditions; however, still have biologically relevant functions in the cell. A significant fraction, more than 1/3rd, of eukaryotic proteins are fully or partially disordered. While IDPs in general perform critical functions in the cell, some of them are also closely related with serious human diseases, including Alzheimer's, Parkinson's, and type II diabetes. It is quite difficult to characterize the heterogeneous disordered ensemble of IDPs with available experimental techniques, and accurate all-atom simulations can be of significant help in interpreting and completing the experimental data. In this poster, I'll highlight our ongoing research aimed at addressing following important questions related to the fundamental properties of disordered proteins:

i) How do protein sequence and temperature affect the properties of IDPs and unfolded proteins? To address this, we simulate five different protein sequences in aqueous solution for the systematic investigation of their characteristic polymer properties. We show that the sequence composition not only encodes properties at room temperature, but also determines their temperature-dependent characteristics.

ii) Where does the disease relevance of IDPs originate? To answer this question we simulate four different mutants of an IDP, Islet Amyloid Polypeptide (IAPP). While human IAPP forms amyloid aggregates in type II diabetes patients, other three mutants have different aggregation characteristics. We identify helix-promoting regions of aggregation prone IAPP monomers, which are potentially responsible for its amyloidogenic, i.e., disease related characteristics (1).

iii) How can we connect our simulations with the experiments? We parameterize a frequently used fluorophore pair of Forster resonance energy transfer (FRET) experiments, which is a useful experimental technique frequently used in IDP research. We have simulated several different unfolded proteins both in the absence and presence of the fluorophores. We find that the most structural properties of unfolded proteins remain unaffected in the presence of FRET chromophores. Our simulations also verify a commonly made critical assumption in FRET experiments, which is hard to justify otherwise (2).

1. Miller, Cayla, Gül H. Zerze, and Jeetain Mittal. "Molecular simulations indicate marked differences in the structure of amylin mutants, correlated with known aggregation propensity." J. *Phys. Chem. B* 117.50 (2013): 16066-16075.

2. Zerze, Gül H., Robert B. Best, and Jeetain Mittal. "Modest Influence of FRET Chromophores on the Properties of Unfolded Proteins." *Biophys. J.* 107.7 (2014): 1654-1660.

433407 Molecular Dynamics Simulations Study of Poly(p-phenylene oxide) Based Polymer Membrane for Alkaline Fuel Cells

Abstract:

Molecular Dynamics Simulations Study of Poly(p-phenylene oxide) Based Polymer Membrane for Alkaline Fuel Cells

Hongchao Pan, Justin Hooper, Dmitry Bedrov

ABSTRACT: To study thermophysical, structural, and transport properties of anion-conductive polymer membrane utilized in alkaline fuel cell applications, atomistic molecular dynamics (MD) simulations employing a polarizable force field have been conducted. The force field was parameterized to fit data from *ab initio* calculations for binding and conformational energies and optimal geometries. A series of systems comprised of p-phenylene oxide dimers and oligomers with tri-methyl ammonium side groups have been simulated in aqueous solutions with hydroxide anion at variety of solution composition (anion-water molar ratios of 1:10, 1:17, and 1:500). Detailed analysis of polymer conformational distributions and dynamics, spatial distribution of water and anions and their dynamics in the polymer matrix were correlated with the underlying heterogeneous structure of membrane. The influence of the length of the alkyl tethers attached to the ammonium-based cation side groups have been also investigated.

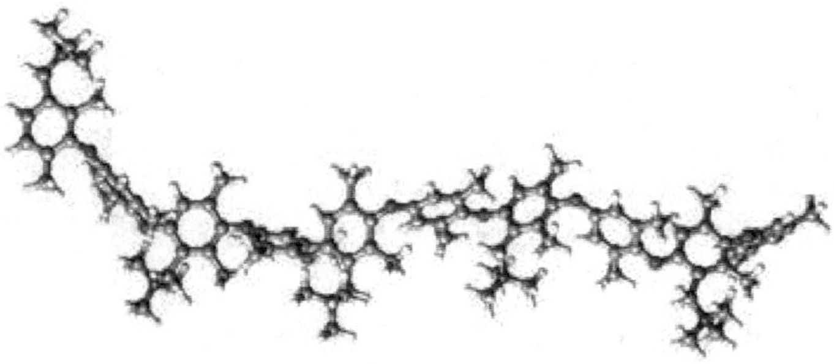

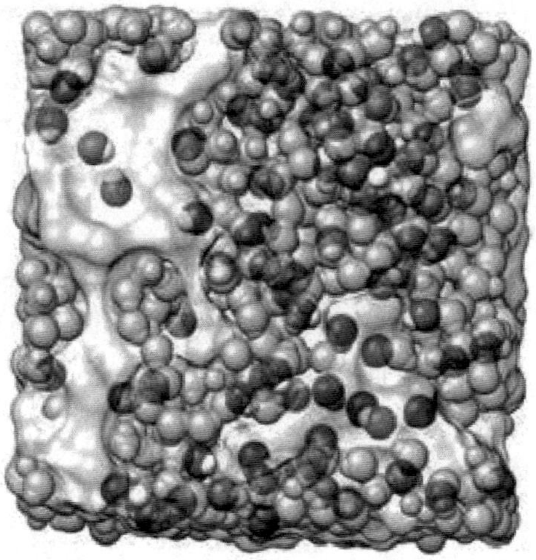

433994 Coarse-Grained Thermodynamic Models for Self and Directed Assembly of Small Ensembles of Colloidal Particles

Abstract:

Self and directed assembly of colloidal particles into crystalline objects is a topic of technological interest, as a route to produce photonic crystals and other meta-materials. Such assembly problems are also of fundamental scientific interest because they involve thermodynamically small systems, with number of particles (10 – 1000) that is far below the bulk limit. Robust methods for controlling the assembly of these crystals would require reduced dimension process-models that link the particle-level dynamics of the colloids to the actuator states. In this poster presentation, we describe the building of such models for two thermodynamically small systems comprising 10-100 micron sized silica particles in aqueous solution.

The first system comprises 32 particles, which interact via a temperature-tunable depletion interaction potential. This system shows interesting phase behavior when the pair interaction strength is changed by a few kT. The second system contains 210 particles and their assembly is controlled by an externally applied electric field. This system shows interesting features like the formation and annealing of grain boundaries during the assembly process as the magnitude of the applied voltage is changed.

We model the assembly process for both these systems using coarse-grained representations, based on the Fokker-Planck equation, which can capture both the dynamics and the equilibrium properties of these small clusters. We use diffusion maps (DMaps), a machine learning technique to identify the slow, low-dimensional manifolds in these systems. The DMap coordinates are correlated against a set of candidate order parameters (OPs) to identify a suitable choice of observables. The DMap technique is sensitive to the nature of defects observed in these two systems and this is manifest in the correlations with OPs. We construct free energy and diffusivity landscapes in the chosen OPs that serve as reduced order models for process control policy maps providing an optimal route to defect-free crystals.

432402 Unraveling the Dissolution Mechanism of Crystalline Cellulose in Alkaline–Urea Solutions

Abstract:

Although it has been recently discovered that alkaline solutions containing urea can dissolve cellulose at relatively low temperatures, the molecular mechanism by which the decrystallization process takes place has not yet been elucidated. Using molecular dynamics simulations, we explore the behavior of cellulose in alkaline–urea solutions, and show that the major effects are caused by an enhancement in the concentration of both the hydroxide group as well as urea at the surface of the cellulose bundle. We compare the results obtained for the alkaline–urea solvent with comparable results obtained for ionic liquid solvents as well as other mixed solvents such as the organocat and organosolv processes.

432697 Insight into the Formation Mechanism of Dimethyl Oxalate By Using the DFT Method

Abstract:

The oxidative coupling reaction of CO and methyl nitride (MN) to dimethyl oxalate (DMO) is an important process for synthesizing ethylene glycol (EG). Palladium (Pd) catalyst is the most effective catalyst for the process. However, using the expensive Pd will lead to the increase in the cost of EG production, thus supported catalysts with ultra-low Pd loading or Pd-free catalysts are needed. Lack in understanding the formation mechanism of DMO and the catalytic mechanism of Pd in the process has significantly hindered the effort. To overcome the challenge, we have employed the density functional theory (DFT) method to study the formation mechanism of DMO at the molecular-electronic level, including the adsorption of every reactant, intermediate and product, MN dissociation, the oxidative coupling of CO, and the formation of final product - DMO.

This work was supported on the National Younger Natural Science Foundation of China (Grant No. 21103120), Program for the Top Young Academic Leaders of Higher Learning Institutions of Shanxi and Wyoming Engineering Initiative in the U.S.

434070 Mie Potentials for Phase Equilibria: Application to Ethers and Sulfides

Abstract:

Optimization of non-bonded parameters in molecular mechanics force fields is a time consuming trial and error process. Methodologies to predict optimized parameters for these models have been developed that use perturbation theory[1], or equations of state, such as SAFT[2, 3], which reduces significantly the range of parameter space that must be explored and the overall computational expense of the optimization. This methodology has been shown to be effective for the development of coarse-grained models[2], and a united atom force field for n-alkanes[4]. Despite their success, however, the application of these methodologies to the development of complex, all-atom force fields that include electrostatic interactions is unclear.

In this work, *ab initio* calculations are used to guide the development of optimized Mie potentials for ethers and sulfides. Potential energy surfaces are determined for neon, argon and Na^+ interacting with propane, dimethyl ether, diethyl ether, dimethyl sulfide, diethyl sulfide, methanol and ethanol at the MP2/aug-cc-PVTZ level. These data are used to provide insight into the transferability of parameters for methyl and methylene groups when placed in different bonding environments, and to predict the relative contribution of each atom to the overall interaction energy. These data are also used to predict the balance between dispersion and electrostatic interactions, and guide the choice of repulsion exponent. Parameters predicted from *ab initio* calculations are refined using grand canonical histogram reweighting Monte Carlo simulations [5-7]. The resulting models predict experimental liquid densities and vapor pressures to within 1% and 10%, respectively.

1. Sans, A., A. Vahid, and J.R. Elliott, *Transferable Intermolecular Potential Models for a Broad Range of Organic Compounds.* Journal of Chemical and Engineering Data, 2014. **59**(10): p. 3069-3079.

2. Mejia, A., C. Herdes, and E.A. Muller, *Force Fields for Coarse-Grained Molecular Simulations from a Corresponding States Correlation.* Industrial & Engineering Chemistry Research, 2014. **53**(10): p. 4131-4141.

3. van Westen, T., T.J.H. Vlugt, and J. Gross, *Determining Force Field Parameters Using a Physically Based Equation of State.* Journal of Physical Chemistry B, 2011. **115**(24): p. 7872-7880.

4. Grosse, A.V., R.C. Wackher, and C.B. Linn, *Physical properties of the alkyl fluorides and a comparison of the alkyl fluorides with the other alkyl halides and with the alkyls of the elements of period II.* Journal of Physical Chemistry, 1940. **44**(3): p. 275-296.

5. Potoff, J.J. and D.A. Bernard-Brunel, *Mie Potentials for Phase Equilibria Calculations: Application to Alkanes and Perfluoroalkanes.* Journal of Physical Chemistry B, 2009. **113**(44): p. 14725-14731.

6. Potoff, J.J. and G. Kamath, *Mie Potentials for Phase Equilibria: Application to Alkenes.* Journal of Chemical and Engineering Data, 2014. **59**(10): p. 3144-3150.

7. http://gomc.eng.wayne.edu.

435055 Self-Assembled Multilayer Stacking of Titanyl Phthalocyanines on Graphene Surface

Abstract:

Self-assembled Multilayer Stacking of Titanyl Phthalocyanines on Graphene Surface

Lalitasri Ravavarapu[±] and Pabitra Choudhury[±]

±Department of Chemical Engineering, New Mexico Tech, Socorro, NM

Abstract:

The creation of flexible and mechanically robust electronic materials with high conductivity and desirable aspect ratio is a major challenge. Conventional electronics is based on solid inorganic materials with very high conductivities but with limited mechanical robustness and flexibility. Conversely, soft organic materials, because of their low conductivity range, limiting their practical implementation in electronics. Recently, the developments of hybrid materials, which have very high conductivity, have also been focused on realizing flexible and conductive microelectronics. However, synthesis, assembly and compatibility of hybrid materials on a flexible substrate remain the critical barriers to develop flexible advanced Nano-electronics. In this study we work with the Titanyl Phthalocyanines (TiOPc) which are non-planar molecules and grow in step flowing mode when deposited on flexible and atomically thin layer of graphene substrate. We used combined Density Functional Theory (DFT) and experiments to understand the growth mechanism of multilayer stacking of TiOPcs on graphene surface by calculating energetics, diffusion barriers and electronic structures of mono/bi/multi-layer TiOPc functionalized graphene substrate.

Acknowledgement:

DFT calculation work was also supported from NSF TeraGrid (XSEDE) resources under allocation number [TG-DMR140131]. Use of the Center for Nanoscale Materials was supported by the US Department of Energy, Office of Science, Office of Basic Energy Sciences, under Contract No. DE-AC02-06CH11357.

435362 Artificial Neural Network Modeling of Alpha-Alkene Polymerization with Zirconocene/MAO Catalyst

Abstract:

Polyolefins, the generic name for synthetic polymers based on ethylene, propylene, and α olefins, have become the world's most common synthetic polymers. The development of new metallocene catalysts allows producing plastics with properties that can be accurately designed; copolymers with controlled molecular weight and monomer sequence distribution, polyolefins with isotactic, syndiotactic, stereoblock or hemiisotactic structures.[1]

A neural network (NN), in the case of artificial neurons called artificial neural network (ANN) is an interconnected group of artificial neurons that uses a mathematical or computational model for information processing based on a connectionistic approach to computation. In most cases an ANN is an adaptive system that changes its structure based on external or internal information that flows through the network. NNs are non-linear statistical data modeling tools. The use of neural networks (NNs) has become increasingly popular for applications where the mechanistic description of the interdependence of dependent and independent variables is either unknown or very complex.[2]

In this work, neural network approach is used to investigate the effect of reaction variables for slurry phase homopolymerization of propene with $Me_2Si(2\text{-}Me\text{-}Ind)_2ZrCl_2$-MAO catalytic system. The experimental data were taken from Palza *et al.*[3] The effect of temperature, propylene concentration (pressure), cocatalyst-catalyst ratio, and catalyst concentration on catalyst productivity and molecular weight of the polymer has been studied using neural network approach in MATLAB® 7.0 software.[4]

References

[1] Bubeck R.A., *Mater. Sc. and Eng. R: Reports*, **39 :1,** p. 1–28 (2002).

[2] Fernandes F. A. N. and Lona L. M.F., *Brazilian Journal of Chemical Engineering*, **22: 03**, p. 401-418 (2005).

[3] Palza H., Velilla T. and Quijada R., *Polymer-Plastics Technology and Engineering,* **45:1**, p. 85- 94.(2006).

[4] MATLAB® version 7.0.1, 2004, computer software, The MathWorks, Inc., Natick, Massachusetts.

434976 Qnpr (Quantitative Nanostructure Property Relationship) Study for Describing Optical Properties of Plasmon Nanomaterials Using 3D Descriptors

Abstract:

Computer Aided Molecular Design (CAMD) has been used to supplement and guide experimental efforts in various fields. Currently, highly predictive models can be developed as a result of improved computing hardware along with novel methodologies for model development. Decreasing dependency on experimental efforts can also reduce the environmental footprint of companies involved in product design and development.

Nanomaterials are getting significant attention due to their unique properties. Specifically, plasmon particles are the focus of much study due to their interesting optical properties. For example, these particles are being tested for various types of biological detection and cell targeting applications. In recent years, attempts have been made to develop models that relate the nanostructure aspect ratio with the optical properties of such particles. In most cases, nanodescriptors were used to develop these models.

This work aims to elucidate how the use of 3D descriptors can influence this type of model development. Genetic Algorithm (GA) in combination with Artificial Neural Network (ANN) is used extensively in the model development. Specifically, we have investigated various levels of generations of GA coupled with different numbers of hidden layers and neurons of ANN during the model identification. Both internal and external validation was used to check the predictive capabilities of the model. An analysis of the application domain was also performed to evaluate the expansive nature of the developed model.

Reference:

1. Nikolai G. Khlebtsov, Lev A. Dykman, *Optical properties and biomedical applications of plasmonic nanoparticles*, 2010, Journal of Quantitative Spectroscopy & Radiative, Transfer 111, 1–35.
2. Ran Chen, Yuntao Zhang, Faryad Darabi Sahneh, Caterina M. Scoglio, Wendel Wohlleben, Andrea Haase, Nancy A. Monteiro-Riviere, and Jim E. Riviere, *Nanoparticle Surface Characterization and Clustering through Concentration-Dependent Surface Adsorption Modeling*, 2014, ACSNano, Vol 8, No. 9, 9446–9456.
3. Xin-Rui Xia, Nancy A. Monteiro-Riviere and Jim E. Riviere, *An index for characterization of nanomaterials in biological systems*, 2010, Nature Nanotechnology, VOL 5, 671-675.

440108 Still Looking for the Magic Spot: Dispersing Modified Fe_2O_3 in Lamellar PS-PMMA Diblock Copolymer By Vapor-Annealing Deposition

Abstract:

Polymer-based nanocomposites (PNCs) have gained popularity in the last two decades due to their exciting bulk and surface properties. However, the current bottleneck for the exploitation of the theoretical excellent properties of PNCs is the complete dispersion of the nanoparticles in the matrix and the consequent development of a huge interfacial area. This complete dispersion will allow maximizing the available matrix–particle interphase optimizing then the organic–inorganic interaction, responsible of the enhanced properties of the final material. So, most of the research efforts in this area have been focused on developing rational processing strategies for PNCs and in promoting better matrix–particle interactions. In this scenario, we present our combined experimental/computational approach to the dispersion of modified Fe_2O_3 nanoparticles in a lamellar PS-PMMA copolymeric matrix. Specifically, we synthesized and characterized several of these PNCs at different nanoparticle concentrations; contextually, multiscale molecular simulations were employed to predict their phase behavior.

430426 Diffusion Wavelet-Based Decompositions for Coarse-Graining of Polymer Chains

Abstract:

We present an alternative approach to coarse graining, based on the multiresolution diffusion-wavelet approach to operator compression, which does not require explicit preparation of atomistic-to-coarse-grained mappings. Our diffusion-wavelet method takes as input the topology and sparsity of the bonding structure of a molecular system, and returns as output a hierarchical set of degrees of freedom corresponding to system-specific coarse-grained variables. Importantly, the hierarchical compression provides a clear framework for modeling at many model scales (levels), instead of the traditional two-level approach used in most coarse-graining applications.

Our results show that the resulting hierarchy separates localized modes, such as a single C-C vibrational mode, from larger-scale motions, e.g., long-range concerted backbone vibrational modes. Our approach correctly captures small-scale chemical features, such as the ring structures in cellulose, alkane side chains, or CH_2 units, as well as large-scale features of the backbone. In particular, the new method's finest-scale modes describe degrees of freedom similar to those found in united-atom models and other chemically-defined CG models. Modes at coarser levels describe increasingly large connected portions of the target polymers. For polyethylene and polystyrene, spatial coordinates and their associated forces were compressed by up to two orders of magnitude. The compression in forces is of particular interest as this allows larger timesteps as well as reducing the number of DoFs.

419715 Atomistic Simulation of the Structure and Mechanics of Semictystalline and Heterogeneous Polymer Systems

Abstract:

Atomistic simulation of the structure and mechanics of semicrystalline and heterogeneous polymer systems

N. Lempesis [(1)], P. J. in 't Veld [(2)], G. C. Rutledge [(3)]

(1) Massachusetts Institute of Technology, Department of Chemical Engineering, USA, Email: nikosl@mit.edu

(2) BASF, Polymer Physics, Germany. Email: pieter.intveld@basf.com

(3) Massachusetts Institute of Technology, Department of Chemical Engineering, USA, Email: rutledge@mit.edu

Abstract: It is generally known that polymer melts rarely crystallize completely. Instead, one obtains a heterogeneous, semicrystalline phase composed of crystalline lamellae interspersed with layers of an amorphous matrix, usually at various orientations within the semicrystalline material. It is believed that the interaction of the crystalline and amorphous domains plays a dominant role in determining thermodynamical, dynamical and mechanical properties of the semicrystalline material. We study amorphous poly(tetramethylene oxide) (PTMO) melt, semicrystalline PTMO and crystalline diphenylmethane-4,4'-diisocyanate (MDI) with 1,4-butanediol as chain extender, as auxiliary systems that are closely related to the components of a common thermoplastic polyurethane (TPU) [1]. This work addresses the challenging question of how the existence of two phases (crystalline and amorphous) exhibiting different chemistry, geometry and composition influences the thermodynamical and mechanical behavior of the composite TPU.

Our approach to understand the structure-property relationships of TPUs entails the development and application of theoretical and computational modeling at the atomistic length scale, where the chemistry, structure and geometrical constraints of individual domain types and the interfaces between them are paramount. In particular, it is widely known that in TPUs, phase separation prevails, where hard, organized and soft, amorphous domains coexist, rendering TPUs in a heterogeneous state. The hard, organized regions comprise predominantly crystalline MDI segments, whereas the soft, amorphous regions consist of amorphous PTMO. In order to simulate the behavior of the composite TPU, a set of auxiliary systems (PTMO melt, MDI and PTMO perfect crystals and semicrystalline PTMO) was initially studied. At a later step, these systems were appropriately combined so that the TPU is obtained. By comparing simulations of the PTMO melt and the PTMO/MDI crystals with experimental data, we have validated an appropriate force field that captures well both intra- and intermolecular interactions for both chemical systems (PTMO and MDI). To model the interface between the crystalline and amorphous domains, we have employed a modified version of the Interphase Monte Carlo algorithm [2]. In general, this algorithm allows the creation of a semicrystalline system by using topology-altering Monte Carlo moves at a targeted area of the perfect crystal. By appropriately utilizing this algorithm we have constructed the semicrystalline PTMO and the heterogeneous TPU. Topological features (tails, loops and bridges) observed in these systems in nature are examined from a statistical point of view. In the process of quantifying the effect of the interface between domains, a criterion has been developed for identifying the relevant interface between domains and a methodology was established that identifies the energetically most favorable interface between any two neighboring components in a heterogeneous material. Results on deformation of crystal and semicrystalline PTMO, as well as crystalline MDI to large strain levels are reported, along with some preliminary results on the deformation of the composite TPU.

Acknowledgement: We gratefully acknowledge financial support from BASF for the research project entitled "Multi-scale Theoretical Modelling of Thermoplastic Polyurethanes".

References:

[1] Bonart R., Morbitzer L. and Hentze G., 1969, J. Macromol. Sci. Phys., B3(2), 337-356; Blackwell J., Kenncorwin H. Gardner, 1979, Polymer, 20, 13-17; Blackwell J., Nagarajan M. R and Hoitink T. B., 1981, Polymer, 22, 1534-1539

[2] In 't Veld P. J, Rutledge G. C., 2003, Macromolecules, 36, 7358–7365

421934 Dissipative Particle Dynamics with Diffusion and Reaction: Application to Blood Clotting

Abstract:

Dissipative Particle Dynamics (DPD) is well-understood to be appropriate for the simulation of flow phenomena at the mesoscopic level. Frequently these are strongly affected by events at much smaller scales, e.g. oxygen transport to red blood cells (RBCs).To account for these smaller scale processes we have incorporated diffusion and reaction of multiple species into an extended version of DPD. At the continuum level this is equivalent to the Navier-Stokes equation plus one diffusion-convection equation for each species. Clot initiation and growth at a damage site on a blood vessel wall involves a number of simultaneous processes including: multiple chemical reactions, species diffusion and flow. The chemistry of clot initiation and growth is now understood to be determined by mechanisms involving many reactions among some two-dozen species in dilute solution. Approximate diffusion coefficients in plasma can be estimated from correlations, while bulk and surface reaction-rate constants are available from the literature. Here the extended DPD is used to simulate the initiation and growth of a clot on a vessel wall. The role of blood particulates, i.e. RBCs and platelets, in the clotting process is beings studied by including them separately and together in the simulations.

430074 Accelerating Ring-Polymer Molecular Dynamics Simulation: A Parallel-Replica Dynamics Approach

Abstract:

Classical molecular dynamics (MD) method is a powerful tool for simulating materials on the atomic scale, which can be used to compute a wide range of equilibrium and dynamical properties, such as elastic moduli, diffusivity, free energy barrier and reaction rate. However, classical MD neglects quantum mechanical nuclear zero-point energy and the nuclear tunneling effects. For systems containing light elements at low temperatures, quantum corrections are necessary in order to obtain a quantitative correct description of the system.

While solving the full time-dependent Shrodinger equation is usually limited to small systems, there are several approximate methods that have been developed for dealing with those complex, many-body systems. Among them, the ring-polymer molecular dynamics (RPMD) method is a popular semi-classical approach to include quantum nuclear effect in a classical MD simulation. RPMD is based on the path-integral formalism of statistical mechanics, which yields real-time MD trajectories that preserve the exact quantum Boltzmann distribution. It has been demonstrated in several systems that RPMD method does very well when used to calculated thermal reaction rates.

A long-standing problem is that conventional MD is limited to relatively short simulation timescales, which leaves many important phenomena out of reach. These so-called activated processes are typically characterized by long periods of uneventful vibrational dynamics punctuated by rapid transitions. While very powerful, RPMD suffers from an even more severe timescale problem as trajectories are generated not on a single system, but on a ring polymer where each bead is a complete replica of the system.

In classical MD, accelerated molecular dynamics (AMD) methods address this challenge by concentrating on the sequence of transitions and shortening the inter-transition intervals. AMD methods require no apriori knowledge of the possible transitions, but to modify the dynamics so that transitions occur faster. In the present work, RPMD was coupled with Parallel Replica Dynamics (ParRep) method to accelerate the simulation of systems containing light atoms at low-temperatures. An example system of helium atom diffusing in the bulk Fe, W, and Fe-Cr alloy at various temperatures will be discussed.

431136 Prion Protein Conformational Statistics Via on-the-Fly Free-Energy Parameterization

Abstract:

On-the-fly free-energy parameterization is a collective-variable biasing approach akin to metadynamics and the adaptive biasing force method, with one important distinction: rather than attempting to build the bias pontential during the simulation using history-dependent accumulation of bias increments, it simply accelerates sampling of collective-variable space using temperature-acceleration and accumulates increments of thermodynamic mean forces. These increments are tallied and integrated into potentials of mean force which ultimately converge. Because it does not use a history-dependent bias, on-the-fly parameterization is not subject to the same pathologies arising from hidden variables that plague metadynamics and ABF. We demonstrate this approach's advantages over metadynamics by revisiting the classical case of beta-hairpin unfolding in the prion protein PrPC. We predict that the free-energy minimum conformation is that observed in the crystal structure, in contrast to previous predictions of metadynamics. We confirm our prediction using several different collective variables. We confirm that well-tempered metadynamics also converges, but we observe that the errors in its prediction of the potentials of mean force are larger than those obtained from on-the-fly parameterization for the same amount of computation, and the metadynamics convergence behavior is more strongly dependent on sampling of hidden variables. Although hidden variables remain a central problem in free-energy methods, on-the-fly free-energy parameterization represents a new approach which may be less susceptible to errors arising from hidden variables than more popular and well-established methods.

432144 Protein Dynamics in the Unfolded State with Optimized All-Atom Protein Models

Abstract:

A significant fraction, more than 1/3rd, of the eukaryotic proteins are partially or completely disordered under physiological conditions. Molecular simulations serve as an ideal tool for atomic level characterization of disordered proteins as experimental characterization can be challenging due to their lack of a single fixed conformation. However, a frequently raised concern in all-atom molecular simulations is that unfolded/disordered proteins being too collapsed as compared to the experiments, suggesting that proteins are insufficiently well solvated in simulations using state-of-the-art force fields. Recently developed models with the use of a more accurate water model and with more balanced protein-water interactions address this issue and produce equilibrium dimensions closer to their experimental values. One major point that remained to be addressed is the protein conformational dynamics in the disordered states. One fundamental dynamic event in unfolded peptides is the residual contact formation. Here, we study the rates of end-to-end contact formation in model unfolded peptides), Cys-(Ala-Gly-Gln)$_n$-Trp, of different lengths (n ranging from 1 to 6), for which experimental tryptophan quenching data are also available. We calculate the tryptophan quenching rates and lifetimes of these peptides using Amber03*/TIP3P (1), Amber03w/TIP4P2005 (2), and Amber03ws/TIP4P2005 (3) protein/water force fields in several microsecond long molecular dynamics simulations. Based on the direct comparison of quenching rates and lifetimes between simulation and experiment, we find that Amber03ws/TIP4P2005 is near optimal for quantitative prediction of both diffusion-limited and reaction-limited rates of contact formation. Whereas simulations with Amber03*/TIP3P provide estimates that are at odd with the experimental values. This suggests that the newer optimized proteins force fields not only improve the equilibrium structural properties of disordered proteins, but can also be used for quantitative understanding of conformational dynamics.

1. Best, Robert B., and Gerhard Hummer. "Optimized molecular dynamics force fields applied to the helix−coil transition of polypeptides." *J. Phys. Chem. B* 113.26 (2009): 9004-9015.

2. Best, Robert B., and Jeetain Mittal. "Protein simulations with an optimized water model: cooperative helix formation and temperature-induced unfolded state collapse." *J. Phys. Chem. B* 114.46 (2010): 14916-14923.

3. Best, Robert B., Wenwei Zheng, and Jeetain Mittal. "Balanced Protein–Water Interactions Improve Properties of Disordered Proteins and Non-Specific Protein Association." *J. Chem. Theory Comput.* 10.11 (2014): 5113-5124.

434344 An Integrated Framework Advancing Membrane Protein Modeling and Design

Abstract:

Membrane proteins are critical functional molecules in the human body, constituting more than 30% of open reading frames in the human genome. Unfortunately, a myriad of difficulties in overexpression and reconstitution into membrane mimetics severely limit our ability to determine their structures. Computational tools are therefore instrumental to membrane protein structure prediction, consequently increasing our understanding of membrane protein function and their role in disease. Here, we describe a general framework facilitating membrane protein modeling and design that combines the scientific principles for membrane protein modeling with the flexible software architecture of Rosetta3. This new framework, called RosettaMP, provides a general membrane representation that interfaces with scoring, conformational sampling, and mutation routines that can be easily combined to create new protocols. To demonstrate the capabilities of this implementation, we developed four proof-of-concept applications for (1) prediction of free energy changes upon mutation; (2) high-resolution structural refinement; (3) protein-protein docking; and (4) assembly of symmetric protein complexes, all in the membrane environment. Preliminary data show that these algorithms can produce meaningful scores and structures. The data also suggest needed improvements to both sampling routines and score functions. Importantly, the applications collectively demonstrate the potential of combining the flexible nature of RosettaMP with the power of Rosetta algorithms to facilitate membrane protein modeling and design.

430985 Mapped Averaging: Reformulation of Ensemble Averages for High-Precision, High-Efficiency Calculation of Properties By Molecular Simulation

Abstract:

We present a general approach that splits thermodynamic derivatives into a known but approximate contribution, plus ensemble averages that quantify deviation from the approximate behavior. Then, direct measurement of the deviations by molecular simulation can be performed without being degraded by noise generated by behavior that is already well characterized. The scheme is demonstrated in several applications, including calculation of anharmonic properties of crystalline solids for a variety of model systems, and calculation of the dielectric constant of dipolar molecules. The approach is easily implemented, and yields thermodynamic properties of a given precision with orders of magnitude reduction in computational effort and no loss of accuracy.

414348 Multiscale from Molecular to Continuum: A Hybrid Simulation Method for Multicomponent Systems

Abstract:

Numerous problems in molecular and interfacial physics involve processes spanning many different length scales, a feature that has motivated efforts in recent years to develop hybrid simulation techniques that allow for atomistic treatment of select regions, and a coarse-grained (e.g. continuum) description for other parts of the problem where a high level of detail is not necessary. Previously, we developed a multiscale simulation strategy using a stochastic particle-based technique called "smoothed dissipative particle dynamics" (SDPD). SDPD is a thermodynamically consistent formulation of continuum solvers such as smoothed particle hydrodynamics that features scale-dependent thermal fluctuations. Using our multiscale approach, it is possible to couple a molecular dynamics (MD) region to a hierarchy of SDPD domains, where each one is characterized by a different length scale. This allows for simulations that simultaneously include regions spanning from the atomistic to the non-fluctuating continuum limit.

While there have been a number of hybrid simulation strategies for single-component fluids proposed in the last few years, extending these types of approaches to multicomponent systems remains a major challenge. In this talk, we describe a novel generalization of our multiscale methods to systems that involve one or more dissolved species. First, we develop a new multicomponent formulation of SDPD through a particle discretization of the diffusion equation with fluctuations in the concentration field. Using this approach, we consider several simple systems and describe how to bridge multicomponent SDPD with current techniques for multiscale SDPD simulation. This is achieved by carefully constructing an interface between SDPD fluids with a different resolution such that momentum, mass, and the amount of solute in the system are globally conserved. In addition, this multicomponent SDPD model can be reconciled with our multiscale MD-SDPD coupling methodology. We discuss the construction of interfaces for coupled MD-SDPD simulations, which requires Monte Carlo trials to match chemical potentials of domains with different resolution. Finally, we consider several simple case studies and demonstrate that these approaches reproduce correct thermodynamic properties at equilibrium.

430892 Entropic Depletion Interactions Between Anisotropic Particles Studied Using a Rigorous, Efficient and Parallelizable Monte Carlo Method

Abstract:

In simulations of the depletion interaction between colloidal particles, the depletants, modeled as penetrable hard spheres, are usually represented by explicit particles. We present a novel Monte Carlo method [1] to implicitly take into account the effective, attractive interaction between arbitrarily shaped hard colloids, by integrating out the depletant degrees of freedom in the grand-canonical ensemble. In contrast to earlier, inexact schemes the method is rigorous, highly efficient and parallelizable and has been implemented on GPUs and multi-core CPUs within the HPMC hard-particle Monte Carlo framework [2]. Our algorithm is useful to study e.g. depletion-stabilized crystallization of anisotropic particles as well as the formation of fibrillar structures from dilute solutions of spheroidal colloids. For systems with very dense or small depletants, we improve efficiency through the use of a configurational-bias scheme already employed in earlier works [3]. The method should therefore in particular enable the simulation of short-ranged, attractive and facet-specific interactions between anisotropic particles, which are necessary for the formation of open colloidal structures, driven by entropy.

[1] Glaser, J., Karas, A. and Glotzer, S.C. *preprint*

[2] Anderson, J. Irrgang, M.E., and Glotzer, S.C., *preprint*

[3] Bolhuis, P. and Frenkel, D. J. Chem. Phys 101, pp. 9689 (1994)
Biben T., Bladon P. and Frenkel D. J. Phys. Condens. Matter 8, pp. 10799 (1996)

432302 Easy Transition Path Sampling in the Microcanonical Ensemble

Abstract:

Molecular simulations of chemical reactions, diffusion in solids, protein folding, and nucleation processes require rare events methods. Transition path sampling methods can simulate these processes without disturbing the natural system dynamics.We present a new algorithm for sampling transition paths: permutation shooting. Permutation shooting rigorously preserves the total energy and momentum of the initial trajectory and is simple to implement even for rigid water molecules. Permutation shooting is demonstrated for two systems: (1) vacancy migration in a two-dimensional crystal, and (2) the association of biomolecules in water.

433022 Incorporation of Chemical Equilibria into Dissipative Particle Dynamics

Abstract:

Dissipative particle dynamics (DPD) provides superb computational efficiency in prediction the mesoscale structure of self-assembled systems. This work is an attempt to extend DPD to systems with chemical equilibria. In such systems, the chemical structure of functional groups depends on the local environment. For example, in proton-exchange polyelectrolyte membranes dissociation of an acid group is influenced by the hydration of that particular group. We suggest a simulation framework that allows incorporation of reversible chemical reactions into soft-core DPD simulations. The reactivity is mimicked through formation and dissociation of breakable bonds implemented using a short-range attractive potential between specific bead types.

We demonstrate our approach on systems with protonation reactions and proton transfer. To model protonation equilibria, we introduce proton as a separate bead type which interacts with bases (such as water and acid anions) via cut-and-shifted Morse potential. First, we consider aqueous solutions of benzenesulfonic acid. The Morse parameters are determined from the experimental ratio of proton self-diffusion coefficient in a large water bath to the self-diffusion coefficient of water and from the experimental acid dissociation constant. We achieve a very good agreement between DPD simulations and theoretical calculations of the acid degree of dissociation. The model of benzenesulfonic acid dissociation is incorporated into DPD simulations of sulfonated polystyrene of different sulfonation degrees and water content. The system segregates into hydrophobic organic and hydrophilic aqueous subphases. We calculate the self-diffusion of water in the segregated structure and estimate proton conductivity using Nernst equation. The calculated water diffusion is in a quantitative agreement and calculated proton conductivity is in a semi-quantitative agreement with the experiment.

Left: Coarse-grained representation of benzenesulfonic acid with the coarse-graining level of three water molecules per bead. Colored blocks on the atomistic model denote the components for the respective DPD beads. C-S dimer represents the deprotonated acid ion, where the equilibrium distance and the cutoff of the associative harmonic potential between beads. Right: mesoscale structure of hydrates sulfonated polystyrene at 25% sulfonation level and hydration level of 8 water molecules per sulfonic group. Water is drawn in green, sulfonate groups in yellow, hydrated protons in black and the hydrophobic parts in red.

434724 Toward a New Generation of Reactive Force Fields: Polarizable Reactive Force Field (ReaxFP)

Abstract:

Modern quantum mechanics (QM) methods can accurately describe intermolecular forces in solids. However, QM calculations are computationally expensive and impractical for the description of the dynamics of systems with hundreds of atoms or more and also for more than nanoseconds of dynamics. Thus, to describe the dynamics of much larger systems for much longer time scales such as dynamical interactions between and within macromolecules, proteins, ceramics, semiconductors, and metallic alloys, it is useful to have an analytic potential or force field (FF) that describes how the forces respond to changing interatomic distances. Development of accurate classical molecular dynamics (MD) FFs, reactive and non-reactive, relies upon accurate expressions for non-bond potential energy terms including electrostatic, van der Waals (VdW), and H-bond interactions. The bond energy contributions are well known from experimental studies and ab-initio calculations.

In particular, the description of the electrostatic interactions is of critical importance because reactive FFs change connectivity during the MD runs and thus require updates of assigned atomic charges to match the new electrostatic environment. Therefore, many of the current approaches which use fixed charges fail to accurately describe the charge distribution within molecules. Since QM leads directly to the electron density throughout space, this might seem as a solution, but most FFs describe the electron density as atom centered point charges. Therefore, other ways (e.g. Mulliken charges, Electrostatic Potential charges, Bader charges, etc) must be used to convert electron densities to atom-centered charges. This can add additional uncertainties due to limitations of these methods. In addition, for large systems, such as virtual ligand screening of ligand-protein interactions, we would prefer to avoid QM calculations for millions of molecules in the library for nanoseconds of the simulations.

For this reason, we propose a polarizable reactive FF (ReaxFP) where a Polarizable Charge Equilibration scheme (PQEq) is used for calculation of electrostatic energy. In PQEq model, the charge on each atom is partitioned into a Gaussian-shaped core with positive fixed charge (nucleus and inner electron) and a Gaussian-shaped shell with negative variable charge (valence electrons). Shell charges can flow from atom to atom based on our earlier Charge Equilibration (QEq) scheme. The shells can be displaced from the cores and the primary restoring force between a core and its shell given by the electrostatic interaction between the two charge distributions. The model uses three atomic parameters for each element (bond radius, electronegativity, and hardness) and 2nd and 4th order spring forces to adjust to the proper atomic polarizability for each element. The charge on each atom is obtained by requiring that the electron chemical potential be the same on all atoms. The optimal charges are obtained using pseudodynamics, which reduces the computational cost to O(N), a large improvement over the $O(N^3)$ required for matrix inversion.

To optimize the electrostatic energy parameters (i.e. PQEq parameters) a constrained optimization was performed on a training set containing a variety of structures for the desired atoms and molecules. Optimization of the three atomic parameters is carried out with respect to the reference charges that are obtained from the electrostatic potential (ESP) of accurate ab-initio calculations and also the Mulliken population method. The polarization energies of electric dipole probes computed by quantum mechanics (QM) are used in the next step to optimize the spring force constants of the model.

ReaxFP uses similar functional forms for valence energy terms as in the reactive FF (ReaxFF), which our group at Caltech also has developed. However, all parameters of the ReaxFP including bonded and non-bonded energy terms must be developed for similar training sets. This is easily done in our Genetic Algorithm based Reactive Force Field optimizer (GARFfield) which is a multi-platform, multi-objective parallel hybrid genetic algorithm (GA) / conjugate-gradient (CG) based force field optimization framework. It enables first-principles based force fields prepared from large quantum mechanical data sets.

We have already optimized ReaxFP for several materials of interest, including $Cu_{2-x}Se$ and GaN. $Cu_{2-x}Se$ based materials are in the forefront of the research for thermoelectric materials with both superionic and electronic conducting behavior. GaN is a very hard, mechanically stable wide bandgap semiconductor material with high heat capacity and thermal conductivity suitable for applications such as in solar cells. We use ReaxFP to address the issues related to the highly mobile cations and the polarization effects in $Cu_{2-x}Se$ and to understand the electrical polarization effects at the GaN interfaces which is a key to proper device simulation. By accounting for polarization, ReaxFP provides an improved description of the dynamics of these materials.

434881 Improving the Efficiency of Visited States Approaches to Expanded Ensemble Simulations

Abstract:

Expanded or generalized ensemble methods are a broad class of methods that can be used to overcome sampling difficulties by adding transitions between thermodynamic ensembles. They are similar to replica exchange methods, except that are performed with single simulations rather than a set of parallel simulations, which makes them much more suitable for many computing environments. These additional thermodynamic ensembles can either be ensembles of interest (for example, the same system at different temperatures if temperature dependent effects are of interest) or auxiliary states added to either improve overlap between end states or to improve sampling by introducing states with faster kinetics.

One weakness of these methods is that in order to sample states at all ensembles with reasonable frequency, simulation weights for each of the sub-ensembles must added. These weights must be determined self-consistently, and the choice of algorithm to determine these self-consistent algorithms can significantly affect the results.

We show how approaches such as Wang-Landau and Transition Matrix methods can be generalized when dealing with expanded ensemble systems containing many thermodynamic states. Specifically, we discuss how these can be seen as approximations to optimal multistate free energy estimators such as MBAR (the multistate Bennett's acceptance ratio method) and as the Rao-Blackwellization of the Wang-Landau estimator. This more general multistate approach allows more efficient determination of free energies and other observables over many states simultaneously, especially when the states are arranged in a multidimensional network, with multiple neighbors in each state.

We additionally show how these multistate estimation methods can be combined with Gibbs sampling techniques, a generalization of Metropolis Monte Carlo methods which includes transitions to all possible states. This combination allows efficient sampling and property prediction over hundreds or thousands of thermodynamic states. We look at applications of these combined methods to searching the parameter space of spatially dependent Ising models as well as as well computing the binding free energy of small molecules to simple molecular containers.

405229 Charge-Transfer Dynamics of Light-Harvesting Systems in Complex Solvated Dnvironments

Abstract:

Photo-initiated charge-transfer processes play a central role in natural systems such as human vision and photosynthesis. While researchers have successfully modified these processes to control simple isolated systems, our understanding of photo-initiated mechanisms in realistic and complex environments is still in its infancy. In particular, recent experiments have shown that coarse-grained descriptions of solvent interactions are unable to accurately capture the electron dynamics in even relatively simple systems. These ongoing observations open an entirely new theoretical field of research in light-activated processes in explicit solvent, with the opportunity to deeply understand the real-time electron dynamics between complex interfaces.

To this end, we have developed a new real-time time-dependent density functional tight binding (RT-TDDFTB) approach to calculate the electron dynamics of donor-acceptor complexes with explicit solvent molecules - all treated at the quantum mechanical level. Our approach significantly differs from previous linear-response TD-DFT methods in that we directly propagate the one-electron density matrix in the presence of a non-perturbative external field. Furthermore, and most importantly, our implementation allows us to calculate the electron dynamics of large solvated systems (~10,000 atoms), whereas conventional approaches are computationally limited to only hundreds of atoms. Using this new capability, we are able to understand and rationalize electron-hole recombination effects as a function of solvent polarity and configuration. Furthermore, this new capability gives us mechanistic insight into the electron dynamics of new systems in complex environments with the goal of guiding experiments in the exploration of charge-transfer dynamics driven by time-dependent external fields.

422042 Computer Simulation of Nucleation of the Ionic Liquid [Dmim+][Cl-] in the Bulk and Near Graphitic Surfaces

Abstract:

Computer simulation of nucleation of the ionic liquid [dmim$^+$][Cl$^-$] in the bulk and near graphitic surfaces

Xiaoxia He,[1] Erik E. Santiso[2] and Francisco R. Hung[1]

[1]Cain Department of Chemical Engineering, Louisiana State University, Baton Rouge, LA 70803, U.S.A.
[2]Department of Chemical and Biomolecular Engineering, North Carolina State University, Raleigh, NC 27695, U.S.A.

Solidification of ionic liquids (ILs) is used in the synthesis of optically-active and magnetic nanomaterials based on ILs. However, very little is understood about the mechanism of nucleation of crystal phases, especially for ILs as most of their applications involve liquid phases. Here we studied the nucleation of crystals of the IL 1,3-dimethylimidazolium chloride, [dmim$^+$][Cl$^-$] from its supercooled liquid phase in the bulk; we also examined systems where the liquid phase is in contact with graphitic surfaces. The string method in collective variables[1,2] was used in combination with Markovian milestoning with Voronoi tessellations[2,3] and order parameters for molecular crystals[4] to sketch a minimum free energy path connecting the supercooled liquid and the crystal phases, and to determine the rates involved in the nucleation process. Our homogeneous nucleation results indicate that, at a supercooling of 58 K, the subcooled liquid has to overcome a free energy barrier of the order of 60-70 kcal/mol and form a critical nucleus with an average size of about 3.6 nm, before it reaches the thermodynamically stable crystal phase. A simulated homogeneous nucleation rate of 5.0 × 10^{10} cm^{-3} s^{-1} was obtained for our system, which is in reasonable agreement with experimental and simulation rates for homogeneous nucleation of ice at similar degrees of supercooling. Similar results for the nucleation of crystals of [dmim][Cl] in contact with graphitic surfaces will be presented and discussed. This study represents our first steps in a series of studies aimed at understanding the nucleation and growth of crystals of organic salts near surfaces and inside nanopores.

[1] L. Maragliano, A. Fischer, E. Vanden-Eijnden and G. Ciccotti, "String Method in Collective Variables: Minimum Free Energy Paths and Isocommittor Surfaces", *J. Chem. Phys.* **2006**, *125*, 024106

+ -

[2] V. Ovchinnikov, M. Karplus and E. Vanden-Eijnden, "Free Energy of Conformational Transition Paths in Biomolecules: The String Method and Its Application to Myosin VI" *J. Chem. Phys.* **2011**, *134*, 085103

[3] E. Vanden-Eijnden and M. Venturoli, "Markovian Milestoning with Voronoi Tessellations", *J. Chem. Phys.* **2009**, *130*, 194101

[4] E. E. Santiso and B. L. Trout, "A General Set of Order Parameters for Molecular Crystals", *J. Chem. Phys.* **2011**, *134*, 064109

An Experimental Design for Predicting the Critical Constant Trends of Large *n*-Alkanes with Gibbs Ensemble Monte Carlo

Richard A. Messerly

Thomas A. Knotts IV

W. Vincent Wilding

Design Institute for Physical Properties (DIPPR 801)

Objectives

1. Minimize the statistical uncertainties in predicting $T_c, P_c,$ and ρ_c from GEMC
2. Demonstrate small finite-size effects for larger compounds
3. Resolve conflicting results found in literature
4. Elucidate the long-chain-length trends for $T_c, P_c,$ and ρ_c of *n*-alkanes

Outline

1. Derive D-optimal design for GEMC

2. Present results and conclusions

Discuss finite-size effects

Slide 3/18

Why a D-optimal design?

- Provides optimal values for independent variables (T)
- Minimizes the uncertainties in the parameters obtained from regression of data (ρ_r and ρ_s)
- The parameters are what we are interested in (ρ_c and T_c)

$$\rho_r \equiv \frac{\rho_L + \rho_v}{2} = \rho_c + A(T_c - T)$$

$$\rho_s \equiv \rho_L - \rho_v = B(T_c - T)^\beta$$

Slide 4/18

D-optimal design

- The covariance matrix (V) for parameters (b) is defined as:

$$V(b) \propto \frac{1}{|F^T W^T W F|}$$

where WF is the Jacobian matrix weighted by the standard deviation in the data (σ)

- A D-optimal design minimizes $V(b)$ by maximizing $|F^T W^T W F|$ (the "D" stands for determinant)

Slide 5/18

D-optimal design for straight-line model

$$y = mx + b$$

With equal weighting (constant σ):

$$|F^T W^T W F| = \frac{(x_1 - x_2)^2}{\sigma^2}$$

To maximize $|F^T W^T W F|$, x_1 and x_2 must be as far apart as possible (limited by physical constraints)

$|F^T W^T W F|$

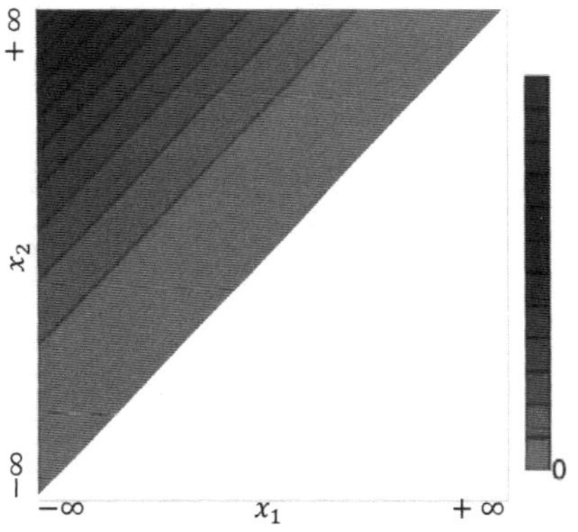

Slide 6/18

GEMC D-optimal with constant σ

- Both equations can be linearized:

$$\rho_r = (\rho_c + AT_c) - AT \equiv r_1 + r_2 T$$

$$\rho_s^{\frac{1}{\beta}} = B^{\frac{1}{\beta}} T_c - B^{\frac{1}{\beta}} T \equiv s_1 + s_2 T$$

- If σ_r and σ_s were constant, the optimal design would be the same as a straight-line model (the extreme values of T)

Slide 7/18

Error model

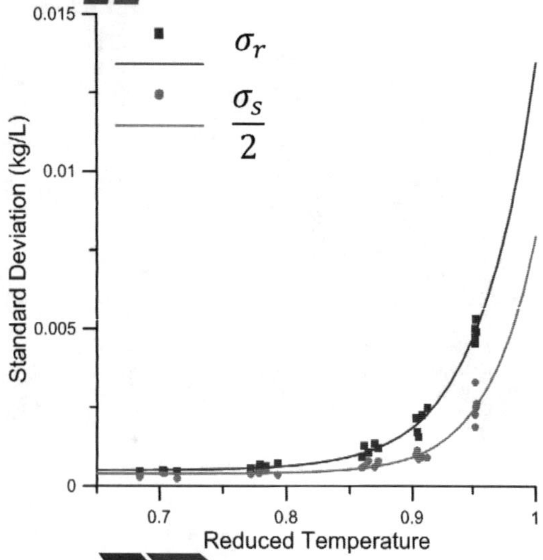

- For non-constant σ, an error model is needed in order to develop an experimental design
- Model developed for C_6, C_8, C_{10}, C_{12}, and C_{16} with simulations of 200 molecules
- Exponentially increasing error has large effect on experimental design

Messerly R. et al., J. Chem. Phys. Vol 143. 2015.

Slide 8/18

GEMC D-optimal design

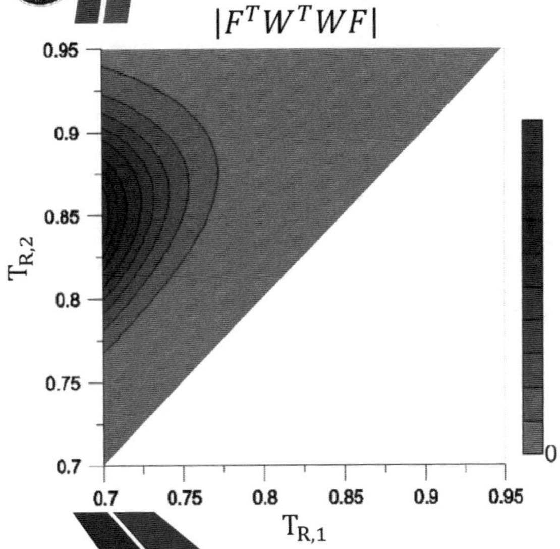

- "Gibbs ensemble simulations suffer from significant size effects, particularly near the critical point."
- "Away from the critical point, there are no significant effects of system size."
- "Finite size effects appear in the vicinity of the critical locus."

Nath et al., J. Chem. Phys. Vol. 108, 1998.
Panagiotopoulos, A. Int. J. Thermophys. Vol. 15, 1994.

Slide 9/18

Comparison with literature

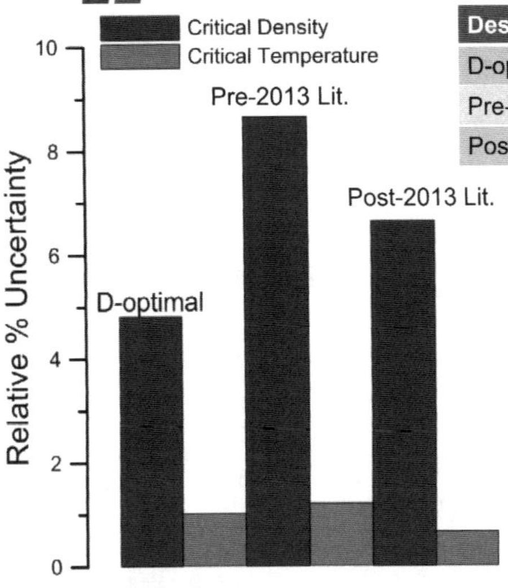

Design	# Temperatures	$T_{R,min}$	$T_{R,max}$	# Replicates
D-optimal	2	0.7	0.85	2
Pre-2013 Lit.	4	0.8	0.95	1
Post-2013 Lit.	4	0.9	0.95	8

Estimates were determined for:

C_{48}
200 molecules

Generated using Monte Carlo statistics and error model

"Relative % Uncertainty" represents two-sided 95% confidence interval

"Critical densities obtained from the rectilinear diameter rule are subject to significantly larger uncertainties than the critical temperatures"

Panagiotopoulos et al., 1994.

Slide 10/18

Simulations in this work

- Three different models: Errington's Exp-6, NERD, and TraPPE
- 200 molecules used (unless specified)
- Equilibration and Production: 200,000 and 300,000 MC cycles
- D-optimal conditions
- 5 replicate simulations at each T_R
- C_{24}, C_{36}, and C_{48}

Errington J. et al., J. Phys. Chem. B. Vol 103, 1999.
Martin M. et al., J. Phys. Chem. B. Vol 102, 1998.

Slide 11/18

Degree of finite-size effects

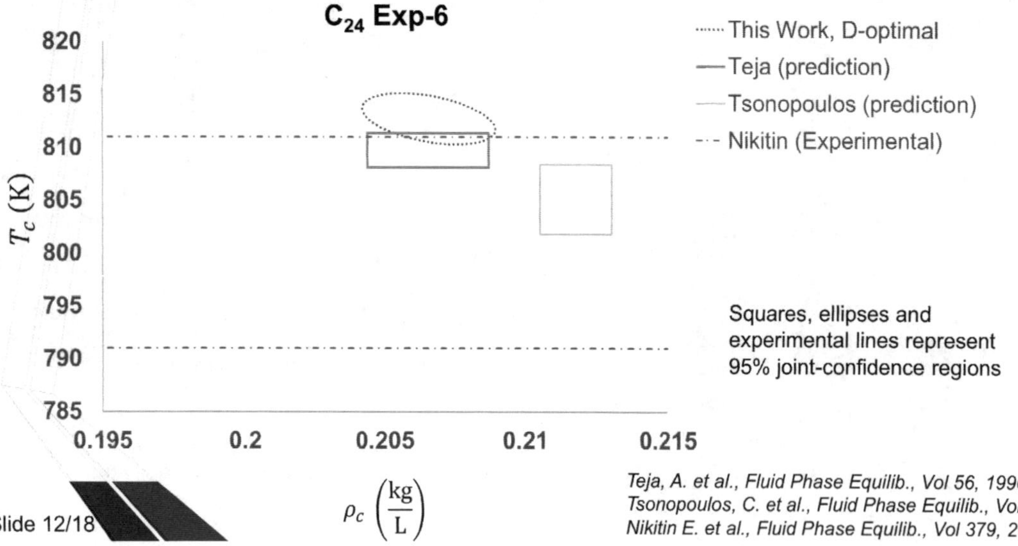

Teja, A. et al., Fluid Phase Equilib., Vol 56, 1990.
Tsonopoulos, C. et al., Fluid Phase Equilib., Vol 83, 1993
Nikitin E. et al., Fluid Phase Equilib., Vol 379, 2014.

Slide 12/18

Degree of finite-size effects

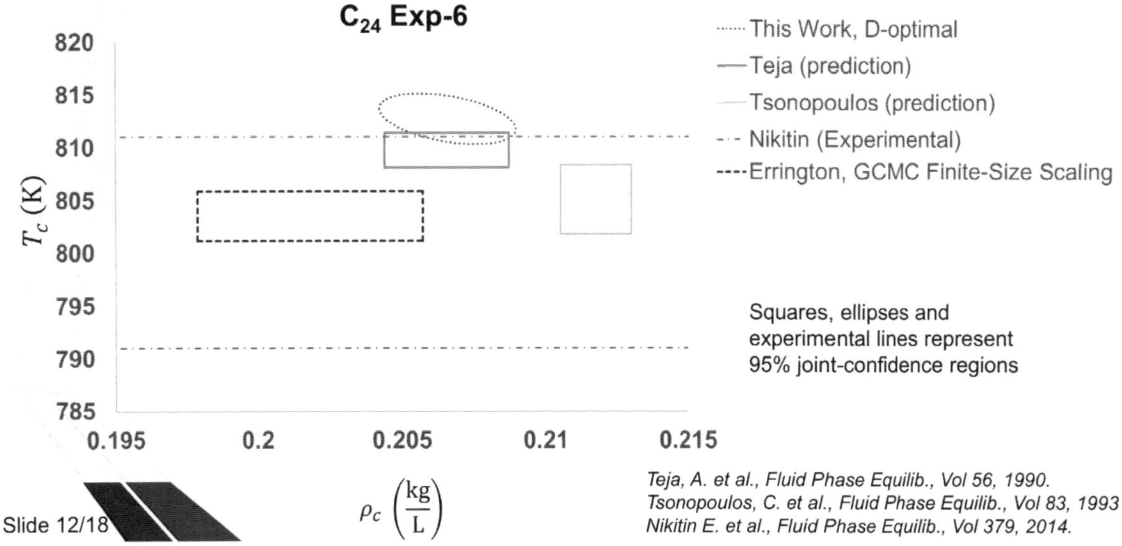

Slide 12/18

Teja, A. et al., Fluid Phase Equilib., Vol 56, 1990.
Tsonopoulos, C. et al., Fluid Phase Equilib., Vol 83, 1993
Nikitin E. et al., Fluid Phase Equilib., Vol 379, 2014.

Degree of finite-size effects

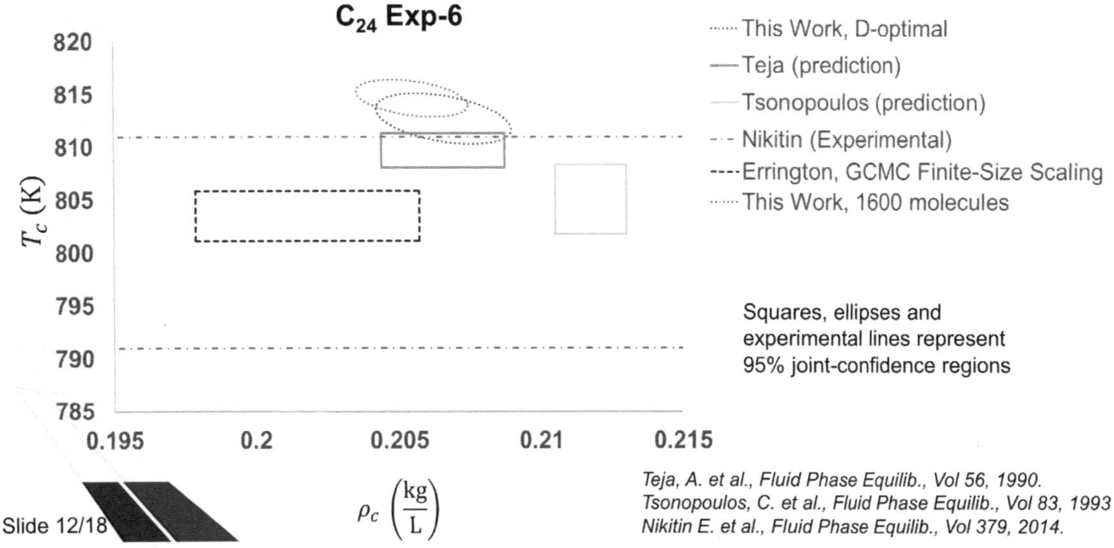

Slide 12/18

Teja, A. et al., Fluid Phase Equilib., Vol 56, 1990.
Tsonopoulos, C. et al., Fluid Phase Equilib., Vol 83, 1993
Nikitin E. et al., Fluid Phase Equilib., Vol 379, 2014.

Degree of finite-size effects (cont.)

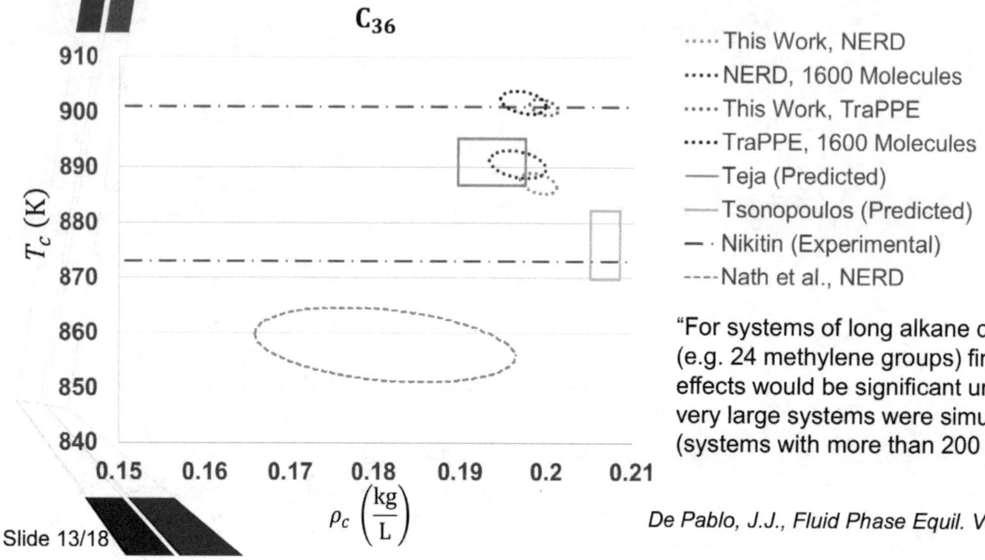

"For systems of long alkane chains (e.g. 24 methylene groups) finite-size effects would be significant unless very large systems were simulated (systems with more than 200 chains)."

De Pablo, J.J., Fluid Phase Equil. Vol. 104, 1995.

Slide 13/18

Resolving T_c trend for *n*-alkanes

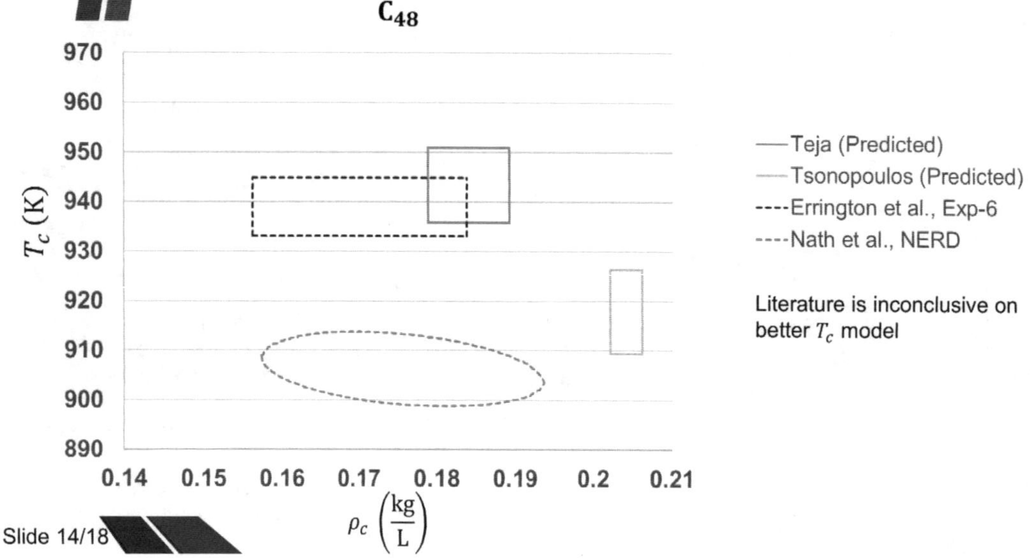

Literature is inconclusive on better T_c model

Slide 14/18

Resolving T_c trend for *n*-alkanes

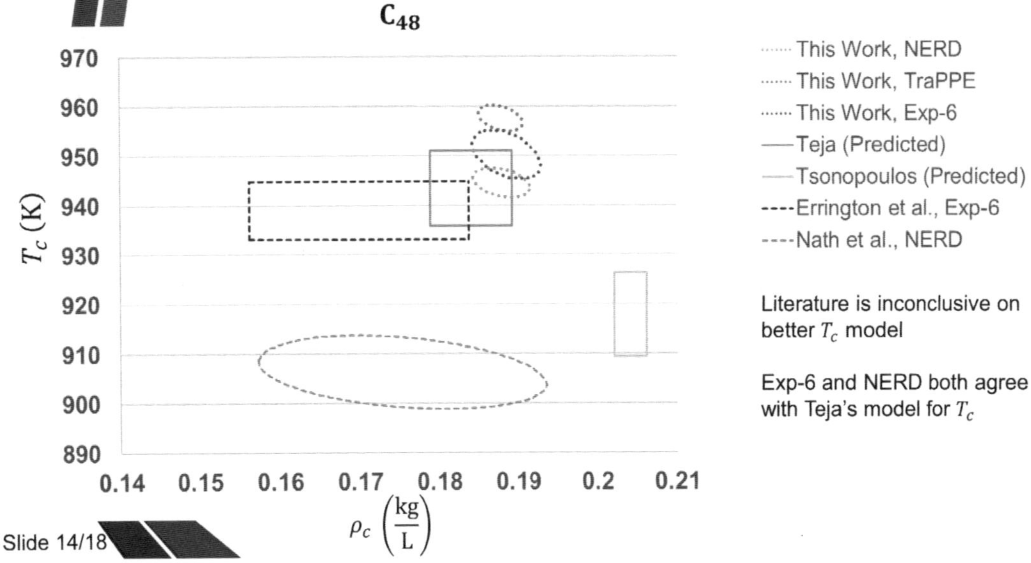

Literature is inconclusive on better T_c model

Exp-6 and NERD both agree with Teja's model for T_c

Slide 14/18

Utilizing D-optimal design to predict P_c

- Vetere's method for predicting P_c with Rackett equation:

$$P_c = \exp\left(\frac{\ln\left(\frac{\rho_c}{\rho_L}\right)}{\left(1 - \frac{T_R}{T_c}\right)^{\frac{2}{7}}} - \ln\left(\frac{M_w}{R_g \rho_c T_c}\right)\right)$$

- Vetere's method requires a low uncertainty in ρ_c and ρ_L

Vetere, A. Chem. Eng. Sci. Vol 44. 1989.

Slide 15/18

D-optimal design results for P_c

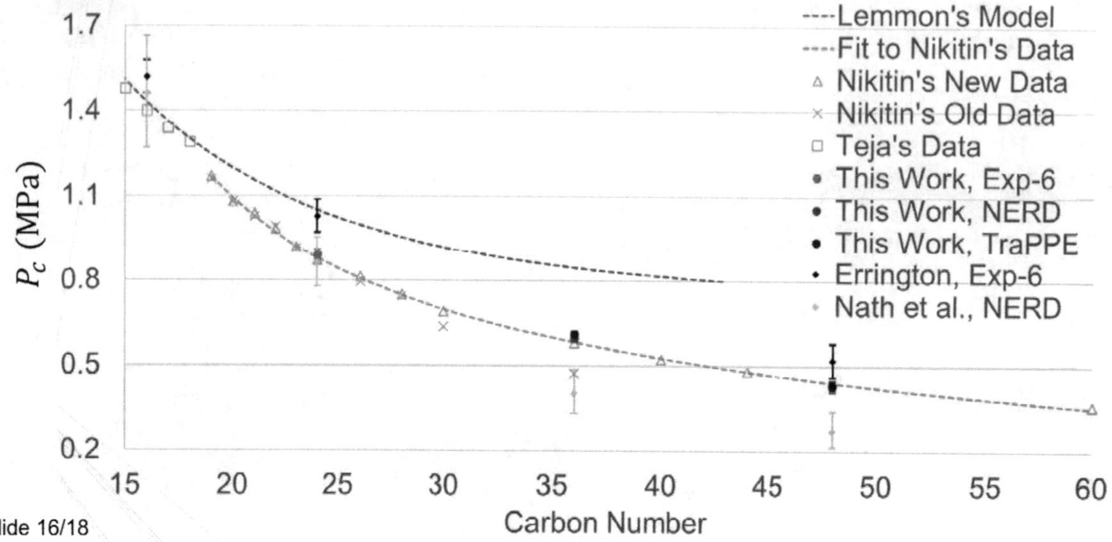

Slide 16/18

Conclusions

- Resolved discrepancy between Errington's Exp-6 and NERD model for long-chain-length trend of T_c for *n*-alkanes
- Finite-size effects are within statistical uncertainty
- Using D-optimal design with Vetere's method yields an uncertainty in P_c that is small enough to elucidate P_c trend for *n*-alkanes
- Nikitin's most recent P_c data are reliable

Slide 17/18

Acknowledgements

DIPPR
Design Institute for Physical Properties
An AIChE Technology Alliance

BYU
FSL

Slide 18/18

426143 The Role of Water in the Growth of NaCl Crystals

Abstract:

Crystal habits are influenced by characteristics of solutes and also by the solvent during crystallization [1]. Numerous studies explore how crystal shape depends on characteristics of the solute and interactions between the solvent and crystal facets [2]. However, several experiments and simulations have implicated desolvation barriers in the rate limiting steps for crystal growth [3-5]. Understanding these kinetic barriers is critical for accurate models of crystal growth. We employ simulation methods for rare events and mechanistic hypothesis testing [6] to understand the ion-desolvation process during attachment to a growing NaCl crystal. We calculate the rate of ion attachment at kink sites to predict growth rates and crystal shapes. We also discuss possible extensions of our findings to other crystals, such as calcium carbonate.

[1] M.A. Lovette, A.R. Browning, D.W. Griffin, J.P. Sizemore, R.C. Snyder, M.F. Doherty, *Ind. Eng. Chem. Res.* **2008**, *47*, 9812-9833.

[2] S.H. Kim, P. Dandekar, M.A. Lovette, M.F. Doherty, *Cryst. Growth Des.* **2014**, *14*, 2460-2467.

[3] P.G. Vekilov, *Cryst. Growth Des.* **2007**, *7*, 2796-2810.

[4] A.G. Stack, P. Raiteri, J.D. Gale, *J. Am. Chem. Soc.* **2012**, *134*, 11-14.

[5] S. Piana, F. Jones, J.D. Gale, *J. Am. Chem. Soc.* **2006**, *128*, 13568-13574.

[6] B. Peters, *Mol. Simulat.* **2010**, *36*, 1265-1281.

429027 Core Level Shifts in Cu-Pd Alloys As a Function of Bulk Composition and Structure

Abstract:

CuPd alloys are important materials in hydrogen purification, where they are used as dense Pd-based separation membranes. Cu is added to impart sulfur tolerance and improved mechanical properties. At intermediate compositions and $T < 873$ K, a BCC alloy (B2) phase occurs, which has superior separation characteristics to those of the FCC phases that form at high Cu and high Pd compositions. Identifying the composition and temperature window where the B2 phase forms is a critical need to enable the design of improved alloys. A composition spread alloy film of Cu and Pd was synthesized. The film was characterized by electron back scatter diffraction and X-ray photoelectron spectroscopy, providing the core level shifts as a function of bulk composition and bulk structure. An anomalous deviation in the Cu core level shift was observed in the composition range $0.33 < x_{Pd} < 0.55$ over which the B2 phase occurs. Density functional theory calculations were used to simulate core level shifts in the FCC and B2 alloy structures. They suggest that the anomalous deviation in core level shift is due to formation of the ordered B2 phase in this composition range.

429709 Finite-Size Corrections to Electrolyte Chemical Potentials Calculated from Molecular Simulations

Abstract:

Finite-size corrections to electrolyte chemical potentials calculated from molecular simulations
Jeff P. Thompson and Isaac C. Sanchez*

McKetta Department of Chemical Engineering, The University of Texas at Austin

*Email: sanchez@che.utexas.edu

Recent studies [1 , 2] have shown increased interest in calculating the concentration dependence of salt chemical potentials (or mean ionic activity coefficients) in electrolyte solutions from molecular simulations, yet a careful analysis of potential errors in the requisite salt solvation free energies due to finite-size effects is necessary to establish the reliability of results. Mean ionic electrolyte chemical potentials computed from free energies of coupling a neutral pair/set of ions to a solution are subject to finite-size errors arising from two major sources: 1) the altered polarization of the solvating medium (pure solvent or electrolyte solution) around an ion in a finite system relative to that in a macroscopic system; and 2) the (macroscopically vanishing) average effective interaction between the distinguished ions whose coupling work is taken as an estimate of the salt excess chemical potential . The former (undersolvation) effect has been discussed extensively in the context of computing single-ion solvation free energies in systems under periodic boundary conditions [3–5]; the effect due to the interaction between distinguished ions has been discussed in the context of the solvation free energy of an ion pair at fixed interion separation [6, 7].

Building on these theoretical investigations, we use continuum electrostatics to estimate the finite-size correction to the solvation free energy of a mobile cation–anion pair in a system with electrostatic interactions treated by Ewald summation. In solutions of high dielectric permittivity, the correction is small relative to the typical magnitude of the salt solvation free energy (at least when, as is often the case, the lattice self-energies of the ions are included in the solvation free energy). Nevertheless, the finite-size error depends on the (thermodynamic state-dependent) permittivity of the solution and may induce non-negligible systematic errors in precise calculations of the concentration dependence of the salt chemical potential. References

[1] Mester, Z.; Panagiotopoulos, A. Z. J. Chem. Phys. 2015, 142, 044507. [2] Moučka, F.; Nezbeda, I.; Smith, W. R. J. Chem. Theory Comput. 2015, 11, 1756–1764. [3] Figueirido, F.; Buono, G. S. D.; Levy, R. M. J. Phys. Chem. B 1997, 101, 5622–5623. [4] Hummer, G.; Pratt, L. R.; García, A. E. J. Chem. Phys. 1997, 107, 9275–9277. [5] Hünenberger, P. H.; McCammon, J. A. J. Chem. Phys. 1999, 110, 1856. [6] Figueirido, F.; Del Buono, G. S.; Levy, R. M. J. Chem. Phys. 1995, 103, 6133. [7] Sakane, S.; Ashbaugh, H. S.; Wood, R. H. J. Phys. Chem. B 1998, 102, 5673–5682.

423696 Rapid Characterization and Validity Quantum Isotope Effect Between Light and Heavy Water through Two-Phase Thermodynamics Model

Abstract:

The quantum isotope effect is important indicated the diverse chemical and physical properties of compounds containing D (Deuterium) or ^{13}C (Carbon-13) etc. The accurate calculations of these properties have a great challenge from molecular simulation. In this work we tried to use same force field parameters to describe the light and heavy water in pure liquid and solid phase behaver based upon MD (Molecular Dynamics) simulation. And then their energy, enthalpy, entropy and heat capacity were analyzed through 2PT (Two-Phase Thermodynamics) model, it is well to exhibit the differences of these properties due to influence of quantum isotope effect for light and heavy water. Owing to the one more neutron causing the mass difference of light and heavy water, so the translational, rotational and vibrational velocity DoS (Density of State) of heavy water is shift to lower frequency relative to light water. Finally these quantum corrections calculation based on the velocity DoS are shown exact phase behaver and physical chemistry properties for light and heavy water. The results of 2PT model have more accuracy relative to the results of PIMD (Path Integral Molecular Dynamics) simulation comparing with experiments using same SPCF2 water model. There is another merit for 2PT model that we just consume a small computational time to calculated large number of atoms system. And then it is easy to extending our method to finish any computation of complex system in future works.

431280 Invited Talk: Data Mining and Machine Learning in Colloidal Science*

Abstract:

There is an enormous design space available to modern colloidal matter, where particle shape and interparticle interactions may be manipulated both in experiment and in computer simulation to span a nearly infinite range of possibilities. Traditional computer simulation methods, such as Monte Carlo and molecular dynamics, have revealed much about the self-assembly behavior of colloidal and patchy particles, and with fast GPU acceleration are producing huge amounts of data on more than 60,000 shapes to date. So what to do with all that data? Although data mining and machine learning techniques such as SVM, neural networks, and genetic and evolutionary algorithms have been used often in bioinformatics and now in the discovery of new "hard" materials, they have yet to be fully exploited in soft matter systems. In this talk we discuss the potential of these powerful techniques in the context of colloid and nanoparticle assembly, and particle design, and present results on a range of systems where data-driven techniques are helping to provide important physical insight.

*with Carl Simon Adorf, Mayank Agrawal, Pablo Damasceno, Paul Dodd, Yina Geng, and Matthew Spellings.

431774 Detecting the True Dimensionality of Complex Dynamical Systems Using Nonlinear Manifold Learning

Abstract:

Reduced macroscale modeling of complex dynamical systems is an integral part of many engineering disciplines. However, positing such reduced models is often difficult, as many systems are described using complex microscale models or via experimental data. Here, as a first step in writing accurate macroscale models for such systems, we use a data-driven methodology to extract the appropriate macroscopic variables with which to describe the dynamics. Because of the diversity and complexity of the potential systems of interest, we use nonlinear manifold learning techniques, which are flexible enough to accommodate a wide variety of structures within a data set, to extract these variables. However, traditional manifold learning techniques are plagued with the issue of "repeated eigendirections," where several of the computed modes parameterize the same direction in the data set. We propose an algorithm for automatically detecting these repeated eigendirections, allowing us to infer the true dimensionality of the data set and construct the most parsimonious reduced model of the underlying dynamical system. We demonstrate our approach using a model which describes cellular chemotaxis.

423147 Local-Environment-Based Distance Metric for Dimensionality Reduction in Colloidal Self-Assembly

Abstract:

Formulating low-dimensional descriptions of complex, dynamical systems is an area of major scientific and technological interest. These low-dimensional descriptions are used to develop computationally tractable models of complex systems and, thus, enable engineering analyses and precise control strategies over these systems. Diffusion mapping is a machine learning technique that can provide dynamically rigorous low-dimensional descriptions for complex systems. This technique has been applied successfully to describe a broad range of physical phenomena from protein folding to colloidal nucleation.

In order to apply diffusion mapping to a particular physical problem the use of a specified distance metric is required. This distance metric should be able to accurately capture the differences between different physical states of the complex physical system under consideration. In this presentation, we will focus on dynamical systems consisting of discrete particles and compare the diffusion mapping results according to three such different distance metrics, namely, the Hausdorff distance, the earth mover's distance, and a novel distance metric based on a description of the average local environment surrounding each individual particle. Each of these three distance metrics is applied to describe the directed self-assembly of a quasi-two-dimensional collection of O(100) colloidal particles. This colloidal system exhibits complex behavior, including order-to-disorder transitions, as well as formation of various micro-structural defects in crystalline assemblies of colloidal particles. This variety in the system structure and physical behavior requires a distance metric which is able to consistently distinguish between different configurations characterized by a broad range of different structural features. This is captured successfully by our local-environment-based description for this system, which is superior to the other distance metrics examined in describing structural diversity.

420140 Near Perfect Prediction of HIV-1 Coreceptor Usage Reveals the Interactions Driving Tropism

Abstract:

In recent years, significant advances in the treatment of HIV-1 have been made, and one class of drugs that has contributed to that success is inhibitors that target chemokine receptors CCR5 and CXCR4, collectively known as the HIV-1 coreceptors (1). These therapeutics, including Maraviroc, are able to circumvent the difficulties of thwarting the quickly mutating HIV-1 by targeting host cell coreceptors, and inhibiting a key interaction with the HIV-1 gp120 V3-loop necessary for entry into host cells. The situation is further complicated by HIV-1 tropism, or the ability of the virus to change the coreceptor used, with the transition from a CCR5-specific (R5) virus to a CXCR4-specific (CXCR4) virus often indicating a progression to advanced stages of infection (2). Therefore, tropism determination is performed in conjunction with coreceptor inhibitors to ensure the success of a treatment regimen. Phenotypic methods, such as the Trofile assay, can be costly with a slow turn-around, so genotypic methods based on sequencing the V3-loop and using bioinformatics methods to predict coreceptor usage are also used (3-6).

New structural data regarding the interactions between the coreceptors and their ligands, including two computationally derived structures of CCR5/CXCR4:V3-loop complexes developed by our group (7, 8), have provided molecular level details of potential interactions driving tropism. However, to date, the specific interactions driving tropism have yet to be identified. To this end, we have developed a multifaceted hybrid approach for elucidating the interactions that drive coreceptor usage, using tools from computational biophysics, structural bioinformatics, and machine learning. Molecular dynamics simulations were used to investigate structural and physicochemical variability of V3-loop:coreceptor complexes. Biophysical insights were converted into structural bioinformatics features using a statistical centroid-centroid force field (9). Nonlinear support vector machines (SVM) were trained to predict coreceptor usage based on the four rules (Net charge, 11/24/25, motif, length)(6), and extracted V3-loop:coreceptor interactions. Finally, a novel non-linear feature selection algorithm was used to narrow down the necessary and sufficient V3-loop:coreceptor interacting pairs.

Of the 624 sets of interactions/rules that were evaluated for predictive utility, the most accurate set achieved a median area under the receiver operator curve (AUC) of 0.981, a median classification accuracy of 0.935, and a median sensitivity of 0.917 at a false positive rate of 5%. We identify a set of only 10 interactions between the V3-loop residues and CCR5/CXCR4 residues, which in combination with the four rules provide a median AUC accuracy of 0.979. An additional 12 interactions, for a total of 22 interactions, are selected when using interactions only and we still attain an AUC accuracy of 0.974. The identified interactions provide new insights into the factors driving HIV-1 tropism, while the resulting highly accurate and low dimensional SVM model represents a novel method for the prediction of coreceptor usage.

1. F. S. Younai, Thirty years of the human immunodeficiency virus epidemic and beyond. International Journal of Oral Science 5, 191–199 (2013).

2. E. A. Berger, P. M. Murphy, J. M. Farber, Chemokine receptors as HIV-1 coreceptors: Roles in viral entry, tropism, and disease. Annual Review of Immunology 17, 657–700 (1999).

3. J. N. Dybowski, D. Heider, D. Hoffmann, Prediction of Co-Receptor Usage of HIV-1 from Genotype. Plos Computational Biology 6 (2010), doi:10.1371/journal.pcbi.1000743.

4. O. Sander et al., Structural descriptors of gp120 V3 loop for the prediction of HIV-1 coreceptor usage. Plos Computational Biology 3, 555–564 (2007).

5. K. Bozek, T. Lengauer, S. Sierra, R. Kaiser, F. S. Domingues, Analysis of Physicochemical and Structural Properties Determining HIV-1 Coreceptor Usage. Plos Computational Biology 9 (2013), doi:10.1371/journal.pcbi.1002977.

6. C. A. Kieslich, D. Shin, A. Lopez de Victoria, G. Gonzalez-Rivera, D. Morikis, A Predictive Model for HIV Type 1 Coreceptor Selectivity. Aids Research and Human Retroviruses 29, 1386–1394 (2013).

7. P. Tamamis, C. A. Floudas, Molecular Recognition of CXCR4 by a Dual Tropic HIV-1 gp120 V3 Loop. Biophysical Journal 105, 1502–1514 (2013).

8. P. Tamamis, C. A. Floudas, Molecular Recognition of CCR5 by an HIV-1 gp120 V3 Loop. Plos One 9 (2014), doi:10.1371/journal.pone.0095767.

9. R. Rajgaria, S. R. McAllister, C. A. Floudas, Distance dependent centroid to centroid force fields using high resolution decoys. Proteins-Structure Function and Bioinformatics 70, 950–970 (2008).

Creating an *in silico* Drug Discovery Pipeline For Faster Drug Discovery

Jonathan Chen, Dept. of Biology

Donald P. Visco Jr., Dept. of Chemical and Biomolecular Engineering

University of Akron

Nov. 11, 2015

The University of Akron

OHIO's POLYTECHNIC UNIVERSITY

1

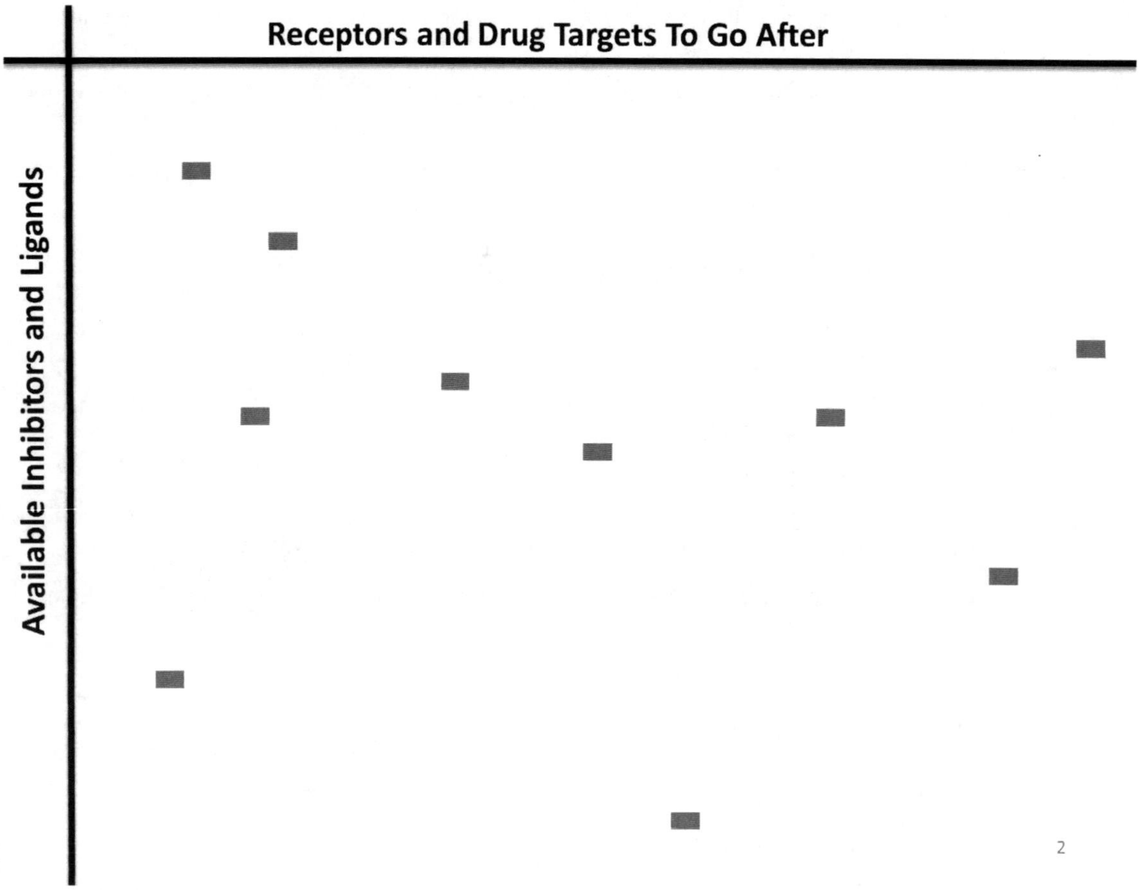

2

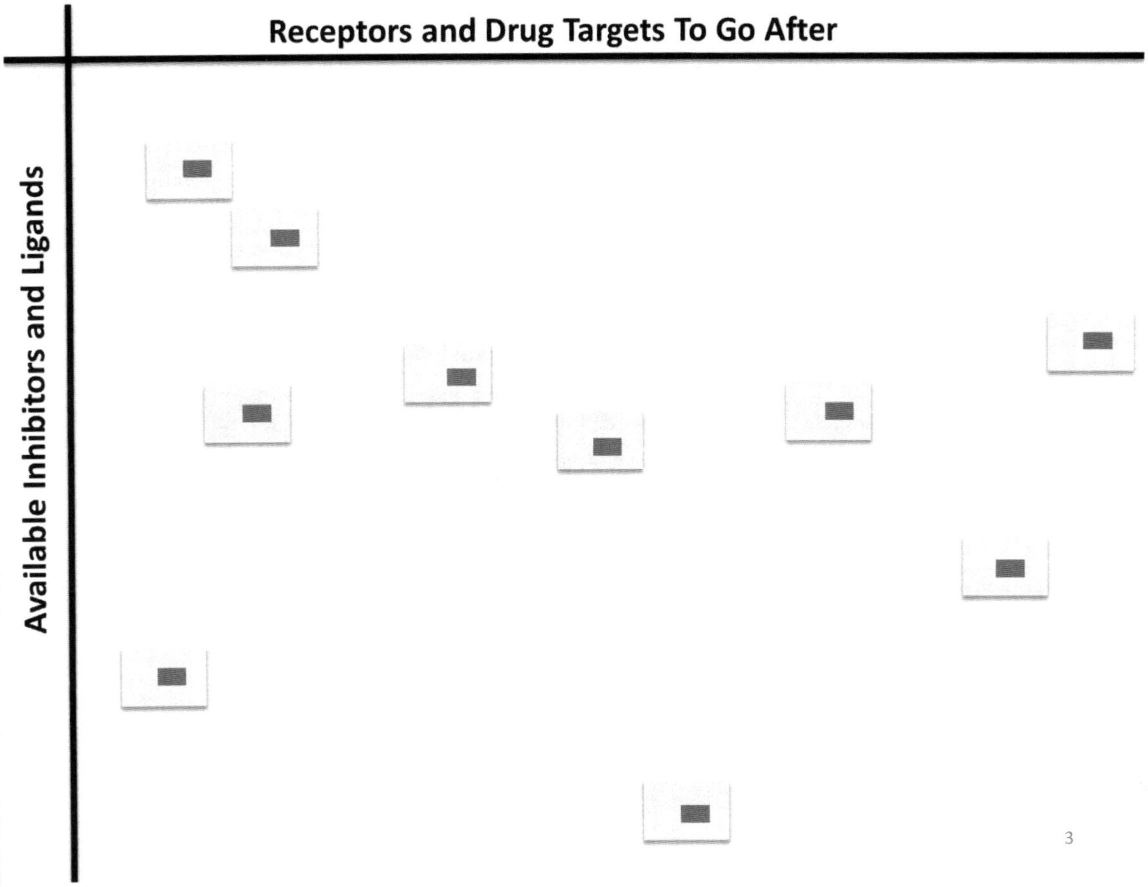

3

In Silico Pipeline

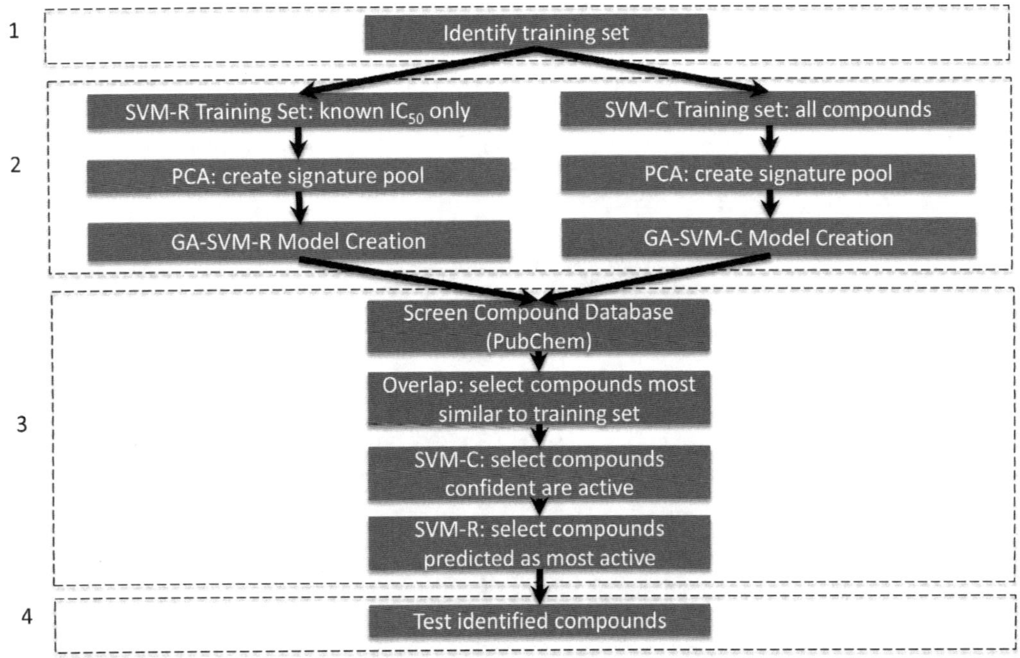

4

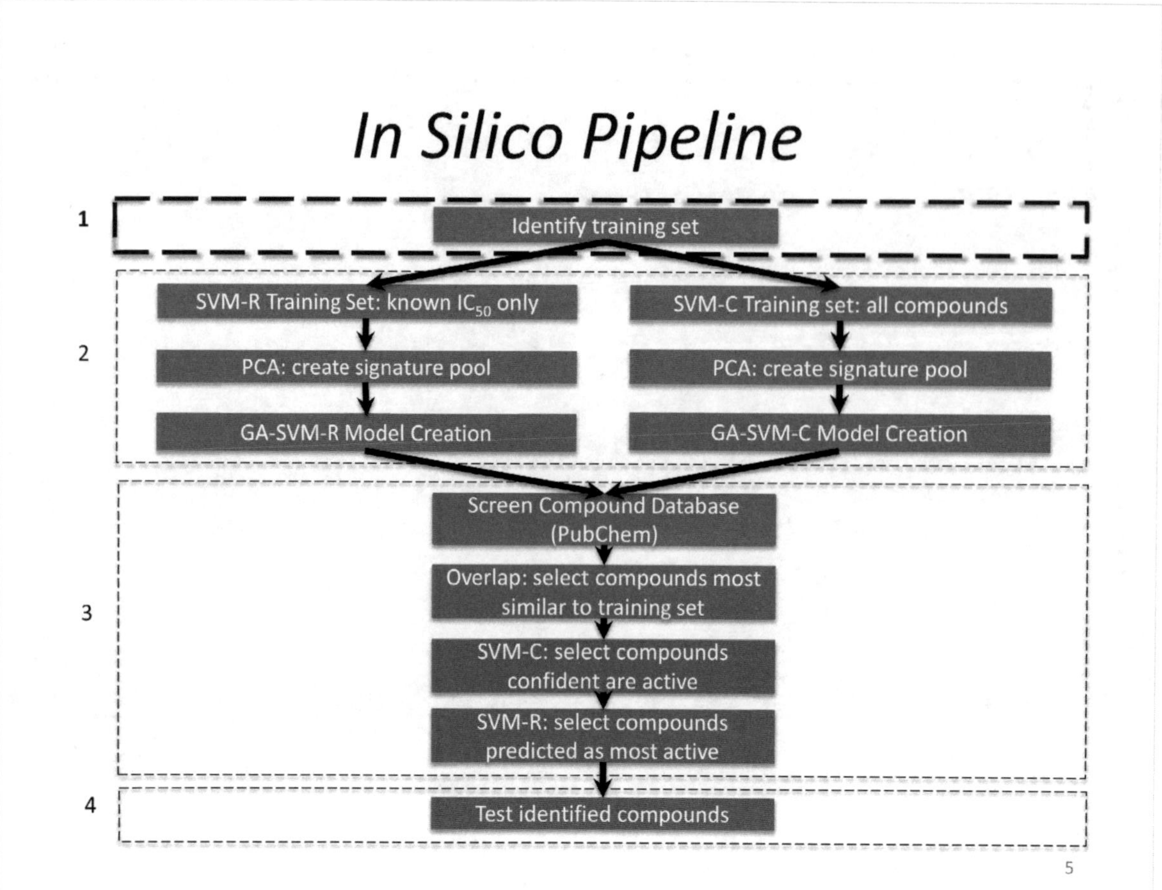
In Silico Pipeline
1
Identify training set
2
SVM-R Training Set: known IC_{50} only
PCA: create signature pool
GA-SVM-R Model Creation
SVM-C Training set: all compounds
PCA: create signature pool
GA-SVM-C Model Creation
3
Screen Compound Database (PubChem)
Overlap: select compounds most similar to training set
SVM-C: select compounds confident are active
SVM-R: select compounds predicted as most active
4
Test identified compounds
5

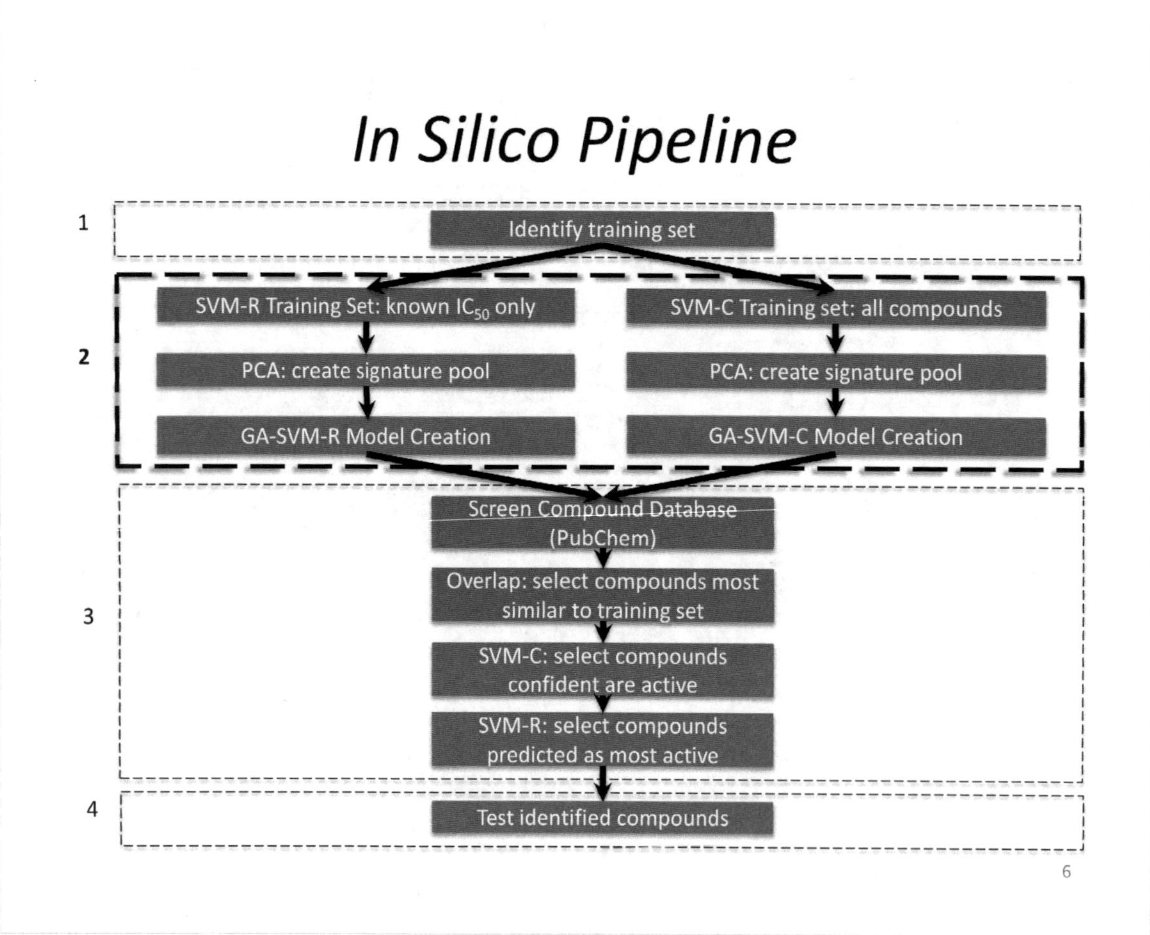
In Silico Pipeline
1
Identify training set
2
SVM-R Training Set: known IC_{50} only
PCA: create signature pool
GA-SVM-R Model Creation
SVM-C Training set: all compounds
PCA: create signature pool
GA-SVM-C Model Creation
3
Screen Compound Database (PubChem)
Overlap: select compounds most similar to training set
SVM-C: select compounds confident are active
SVM-R: select compounds predicted as most active
4
Test identified compounds
6

In Silico Pipeline

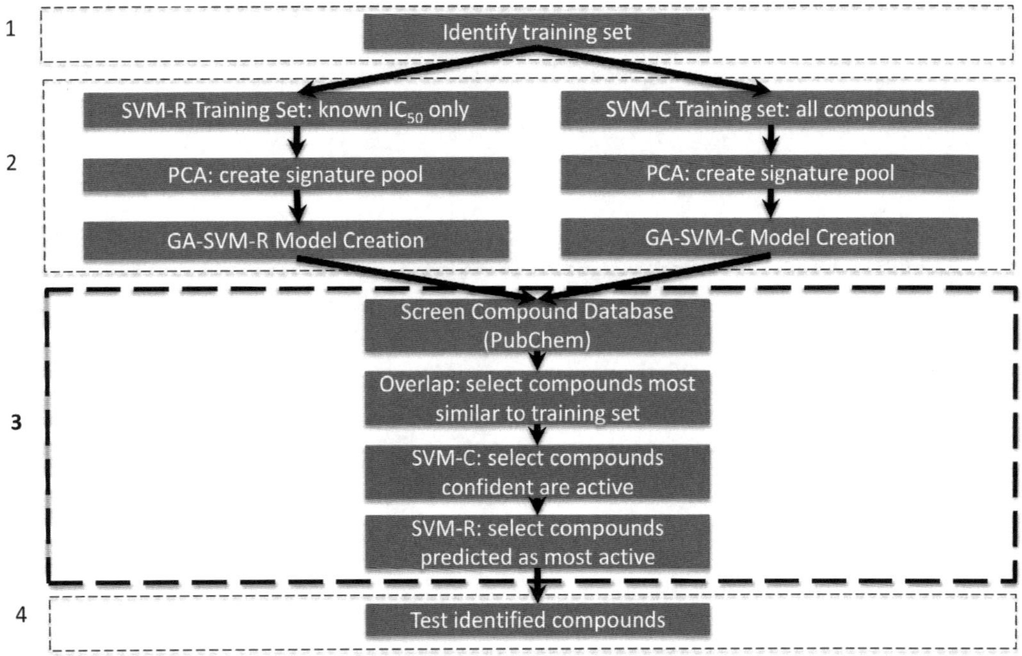

7

In Silico Pipeline

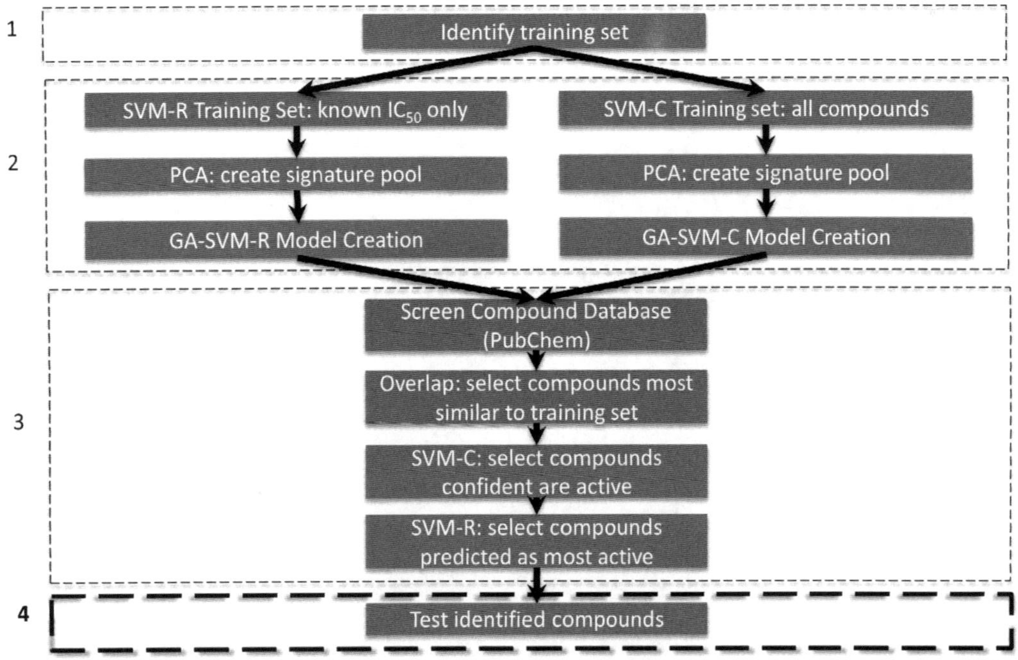

8

In Silico Pipeline

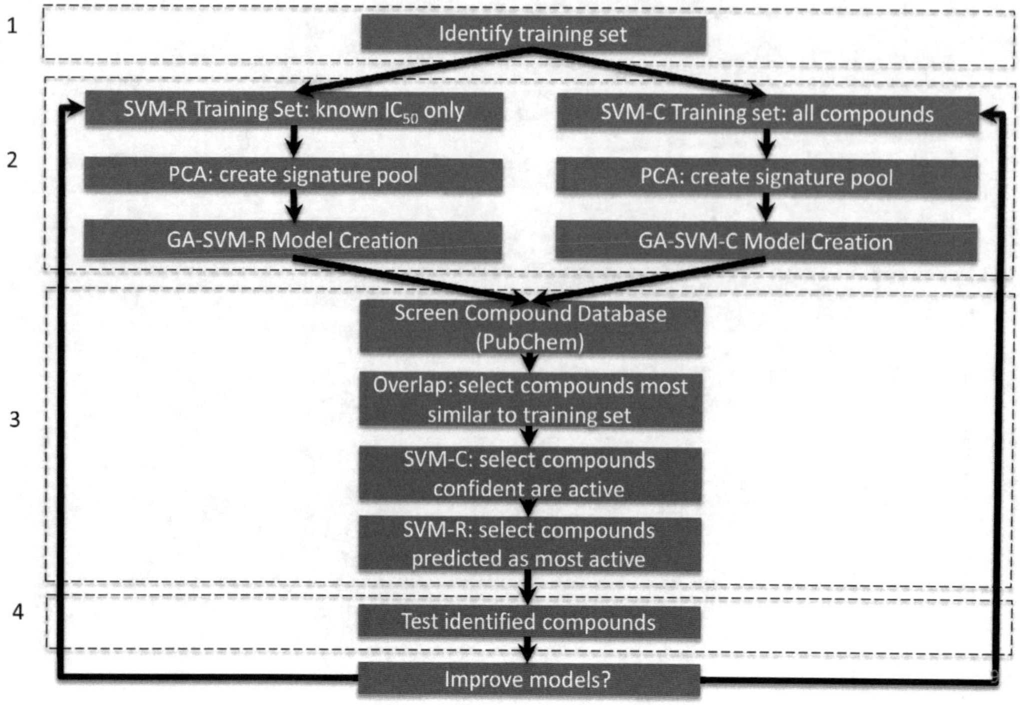

Signature Molecular Descriptor

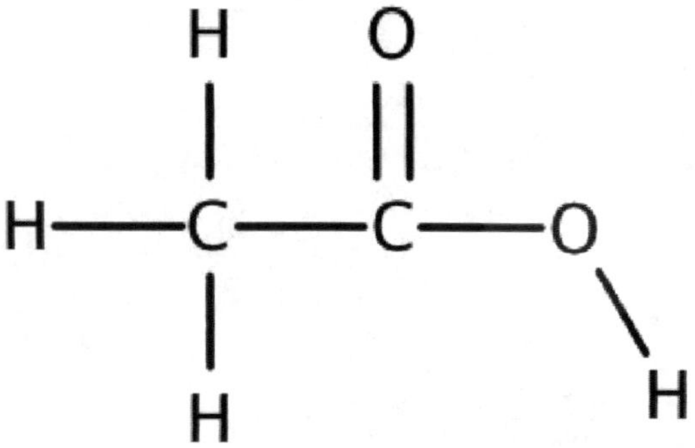

10

Signature Molecular Descriptor

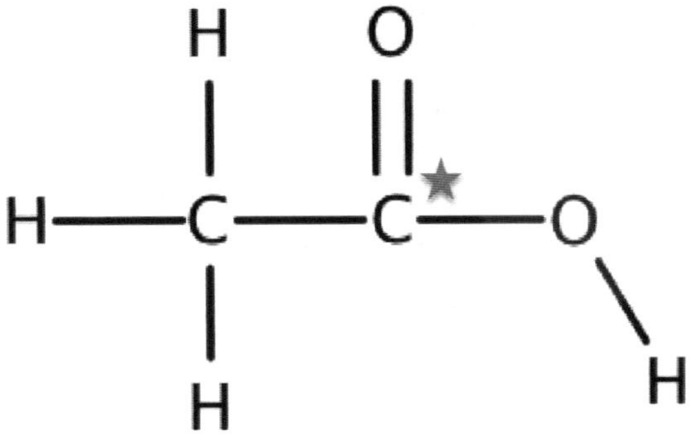

Height=0:
Atomic Signature for C:
C

Molecular Signature:
2 C+ 2O+ 4H

11

Signature Molecular Descriptor

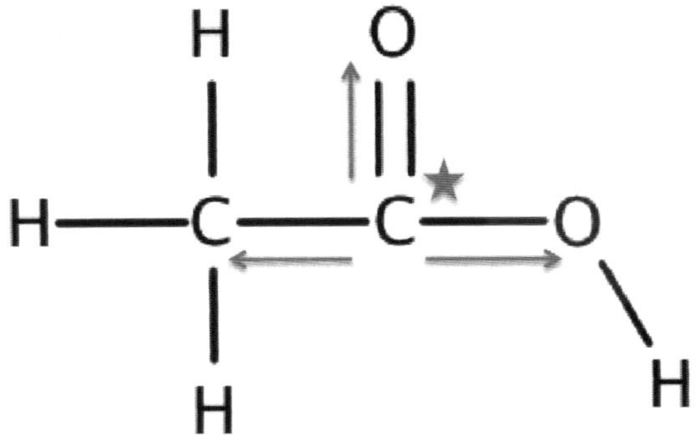

Height=1:
Atomic Signature for C:
C(C,=O,O)

Molecular Signature:
1 C(H,H,H,C)+ 1 C(C,=O,O)
+ 1 O(=C)+ 1 O(C,H)+ 3 H(C)
+ 1 H(O)

Height=0:
Atomic Signature for C:
C

Molecular Signature:
2 C+ 2O+ 4H

12

Signature Molecular Descriptor

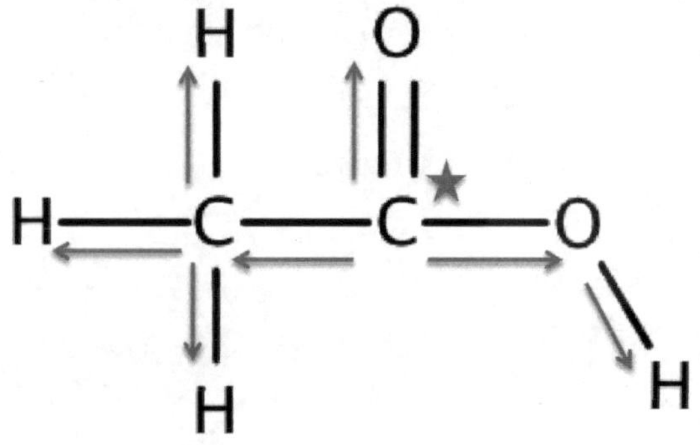

Height=1:
Atomic Signature for C:
C(C,=O,O)

Molecular Signature:
1 C(H,H,H,C)+ 1 C(C,=O,O)
+ 1 O(=C)+ 1 O(C,H)+ 3 H(C)
+ 1 H(O)

Height=0:
Atomic Signature for C:
C

Molecular Signature:
2 C+ 2O+ 4H

Height=2:
Atomic Signature for C:
C(C(H,H,H),=O,O(H))

Molecular Signature:
1 C(H,H,H,C(=O,O))
+ 1 C(C(H,H,H),=O,O(H))
+ 1 O(=C(C,O))+ 1 O(C(=O,C),H)
+ 3 H(C(H,H,C))+ 1 H(O(C))

13

Signature Molecular Descriptor

Faulon, J.-L. Stochastic Generator of Chemical Structure. 1. Application to the Structure Elucidation of Large Molecules. *J Chem Inf Comp Sci* **34**, 1204-1218, doi:10.1021/ci00021a031 (1994).

Faulon, J. L., Visco, D. P., Jr. & Pophale, R. S. The signature molecular descriptor. 1. Using extended valence sequences in QSAR and QSPR studies. *J Chem Inf Comput Sci* **43**, 707-720, doi:10.1021/ci020345w (2003).

Churchwell, C. J. *et al.* The signature molecular descriptor. 3. Inverse-quantitative structure-activity relationship of ICAM-1 inhibitory peptides. *J Mol Graph Model* **22**, 263-273, doi:10.1016/j.jmgm.2003.10.002 (2004).

Weis, D. C., Visco, D. P., Jr. & Faulon, J. L. Data mining PubChem using a support vector machine with the Signature molecular descriptor: classification of factor XIa inhibitors. *J Mol Graph Model* **27**, 466-475, doi:10.1016/j.jmgm.2008.08.004 (2008).

Weis, D. C. & Visco, D. P. Computer-aided molecular design using the Signature molecular descriptor: Application to solvent selection. *Comput Chem Eng* **34**, 1018-1029, doi:DOI 10.1016/j.compchemeng.2009.10.017 (2010).

Li, H., Visco, D. P. & Leipzig, N. D. Confirmation of predicted activity for factor XIa inhibitors from a virtual screening approach. *AIChE Journal* **60**, 2741-2746, doi:10.1002/aic.14508 (2014).

Kayello, H. M., Tadisina, N. K. R., Shlonimskaya, N., Biernacki, J. J. and Visco, D. P. (2014), An Application of Computer-Aided Molecular Design (CAMD) Using the Signature Molecular Descriptor—Part 1. Identification of Surface Tension Reducing Agents and the Search for Shrinkage Reducing Admixtures. J. Am. Ceram. Soc., 97: 365–377. doi:10.1111/jace.12453

Chemmangattuvalappil, N. G. & Eden, M. R. A Novel Methodology for Property-Based Molecular Design Using Multiple Topological Indices. *Ind Eng Chem Res* **52**, 7090-7103, doi:10.1021/ie302516v (2013).

14

In Silico Pipeline

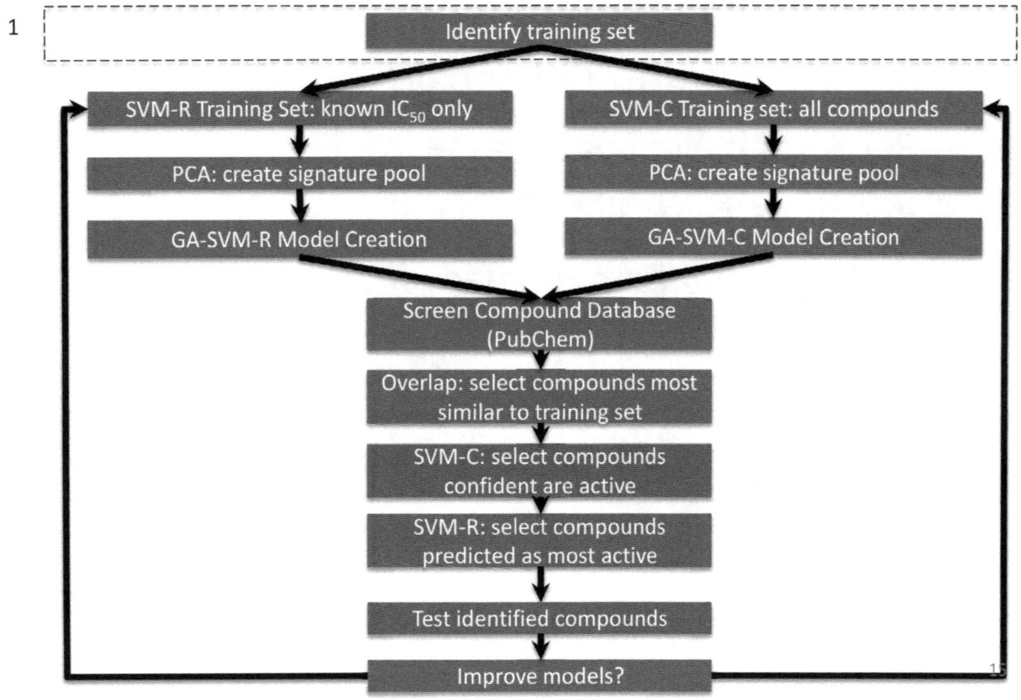

1. Identifying Training Set

- Data to train our models
- More data than ever before
  - NCBI's PubChem Database
  - European Bioinformatics Institute's ChemBL database
  - Proprietary/private database

16

1. Identifying Training Set

- Proof-of-concept: Cathepsin L
  - Receptor in viral pathways
  - Inhibitors = possible anti-viral drugs
  - Dataset: PubChem Bioassay AID 825
    - 48 actives (IC_{50} < 50uM), 50 inactives (IC_{50} >50uM)

Actives

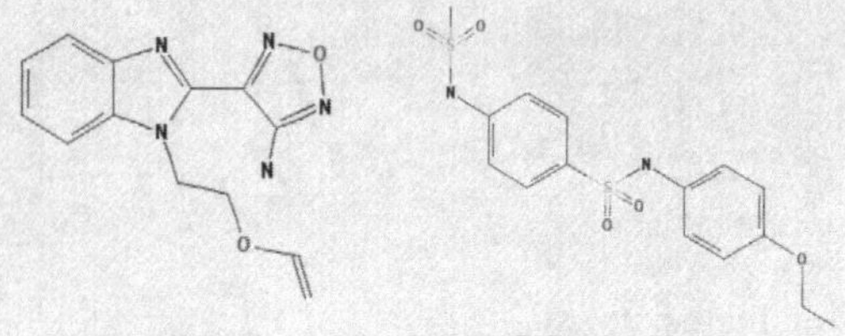

CID 2883004 CID1315081

Inactives

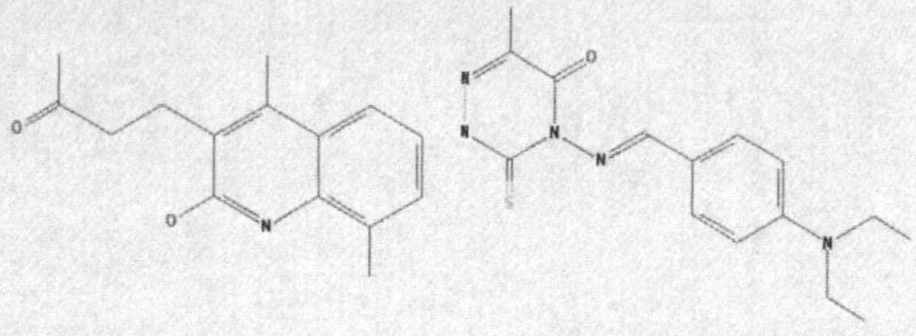

CID 790430 CID 6858457

17

In Silico Pipeline

2

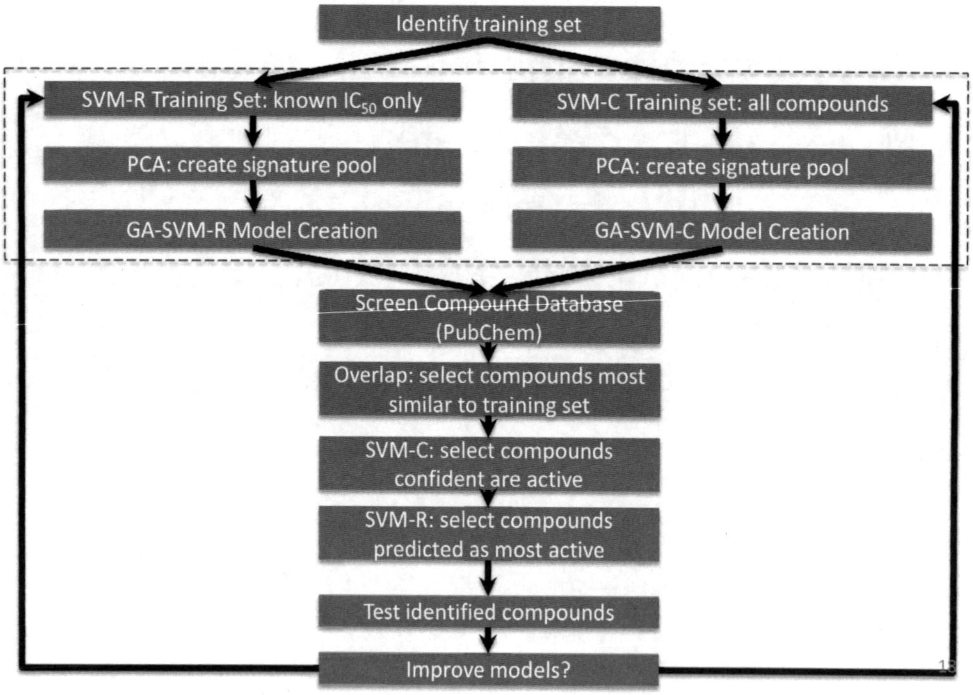

2. Model Training

- Principal Component Analysis-Genetic Algorithm-Support Vector Machine

19

2. Model Training

- Principal Component Analysis-Genetic Algorithm-Support Vector Machine

Signatures
$h^0 = 8$
$h^1 = 127$
$h^2 = 770$

Total = 905

20

2. Model Training

- **Principal Component Analysis**-Genetic Algorithm-Support Vector Machine

Signatures
$h^0 = 8$
$h^1 = 127$
$h^2 = 770$
Total = 905

PCA

Significant signatures: 283

2. Model Training

- Principal Component Analysis-**Genetic Algorithm**-Support Vector Machine

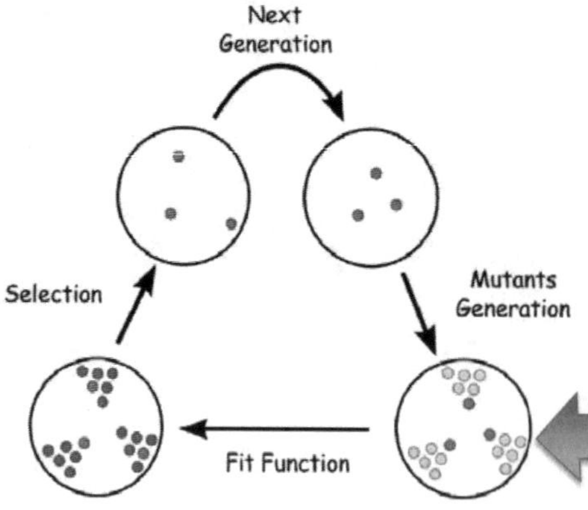

Signatures
$h^0 = 8$
$h^1 = 127$
$h^2 = 770$
Total = 905

PCA

Significant signatures: 283

https://nnpdf.hepforge.org/html/GA.html
http://www.intechopen.com/source/html/18993/media/image8.png

2. Model Training

- Principal Component Analysis-Genetic Algorithm-**Support Vector Machine**

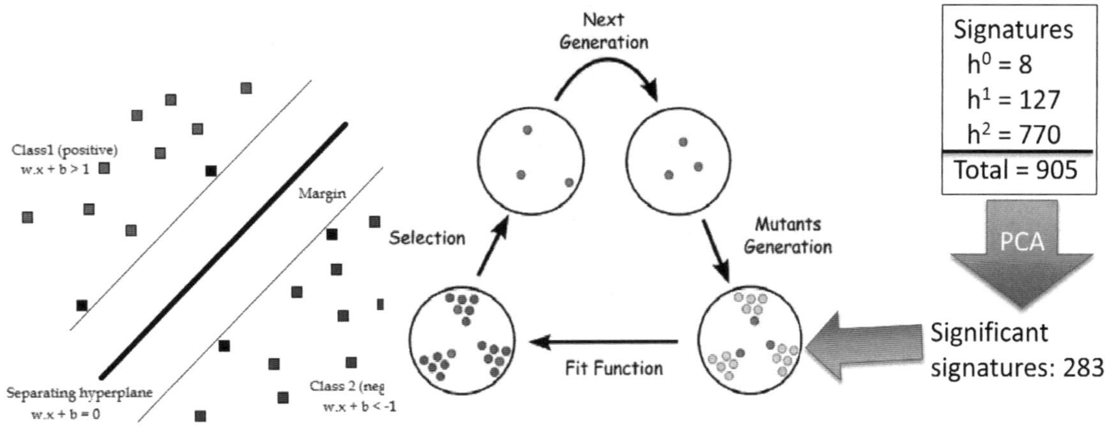

https://nnpdf.hepforge.org/html/GA.html
http://www.intechopen.com/source/html/18993/media/image8.png

23

2. Model Creation

- Proof-of-Concept: Cathepsin L
  - 1st virtual HTS models
    - 1 Classification Model
      - Training Error: 0.020
      - Cross Validation Error: 0.019
    - 1 Regression Model
      - Training Error: 0.071
      - Cross Validation Error: 0.067

24

In Silico Pipeline

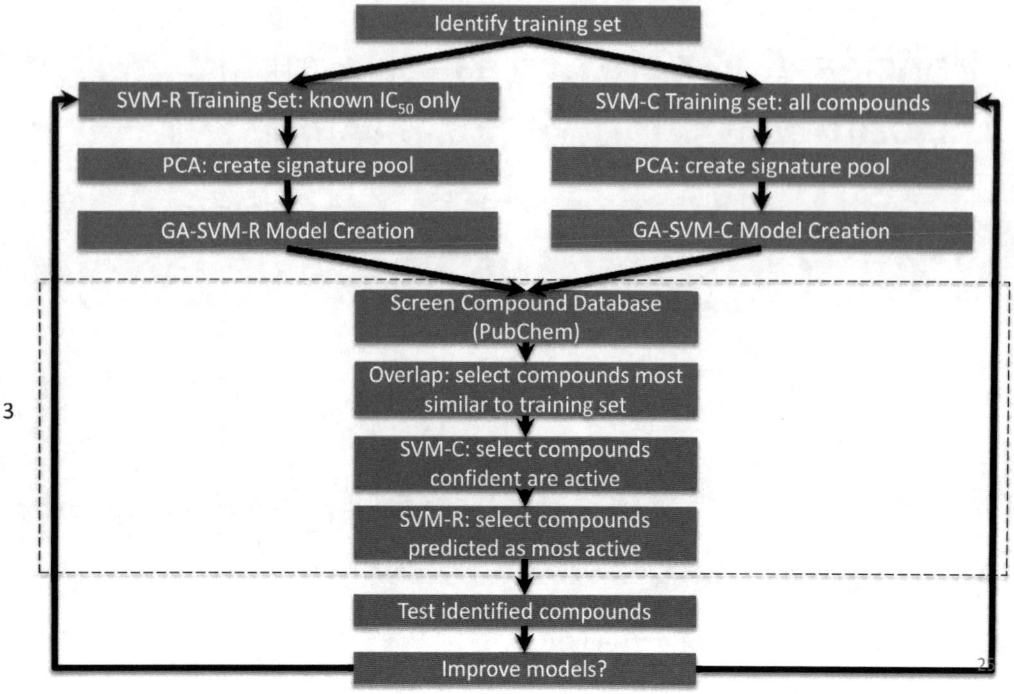

3. Virtual HTS

- Fragment compounds in compound database
- Run compound fragments through models
- Classification and regression selection
  - Classification: qualitative, identify possible candidates
  - Regression: quantitative, identify candidates predicted most active

26

3. Virtual HTS

- Proof-of-Concept: Cathepsin L
  - Screen 72 million compounds with classification and regression model
    - Identified 39 compounds
    - 16/39 commercially available, purchase for testing

27

In Silico Pipeline

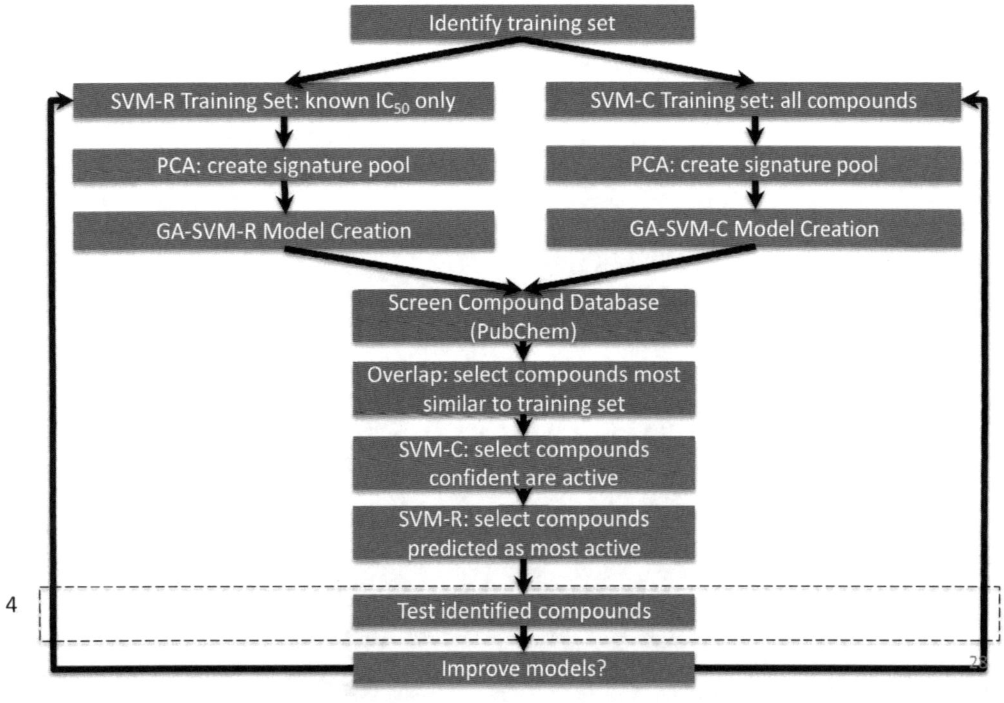

28

4. Test Identified Compounds

- Proof-of-Concept: Cathepsin L
  - Fluorescence Assay
  - 3/16 active (discovery rate 18.75%)

CID 17644424
Median TC: 0.39
Predicted IC_{50}: 0.46uM
Experimental IC_{50}: 20.80uM

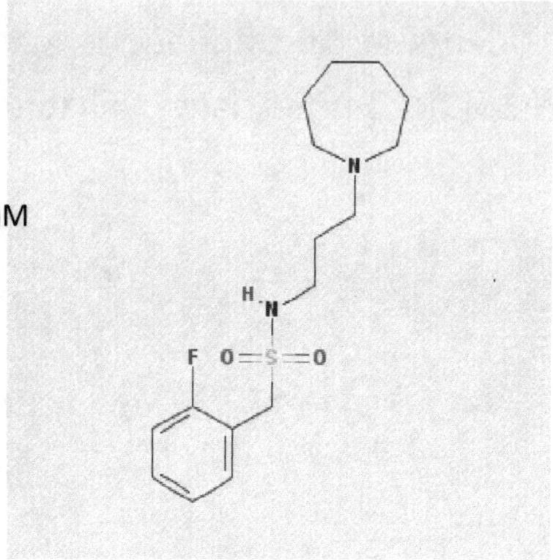

29

In Silico Pipeline

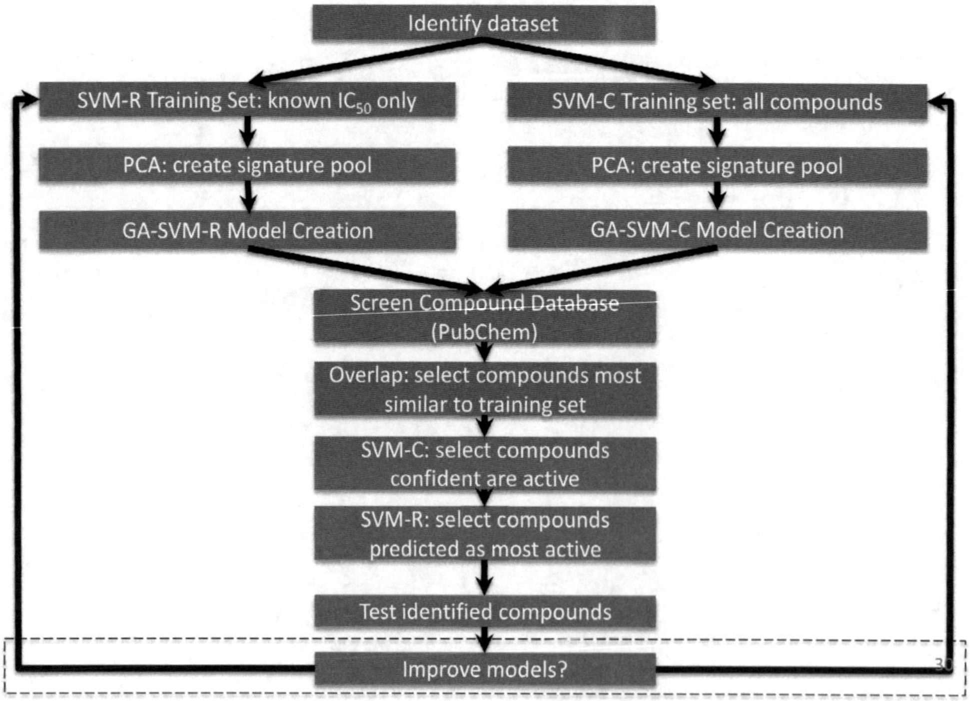

30

Retrain Models

- Proof of Concept: 2nd virtual HTS pass
  - 2 Classification Model
    - Training Error: 0.060
    - Cross Validation Error: 0.204
  - 1 Regression Model
    - Training Error: 0.062
    - Cross Validation Error: 0.299
  - Identified 23 compounds
  - 13/23 commercially available
  - 9/13 active (69.23% discovery rate)

31

Results Discussion (TC=0.968)

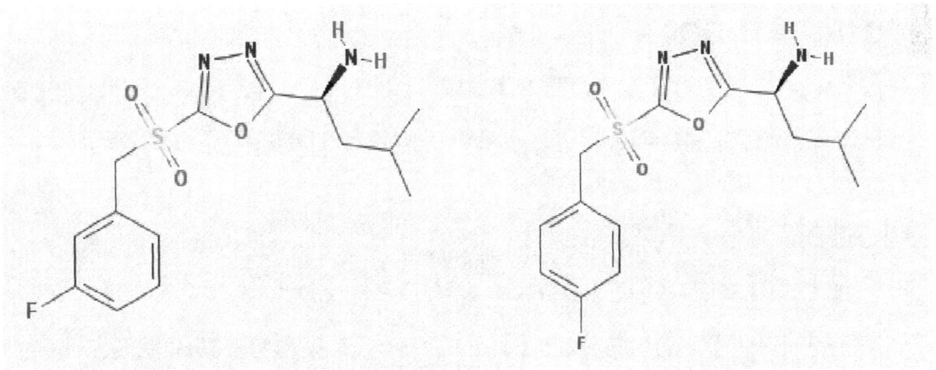

CID 6603229 (training set – AID 825)
Expt. IC_{50}: 4.57uM

CID 6603233 (vHTS)
Expt. IC_{50}: 0.49uM
Pred. IC_{50}: 8.16uM

32

Results Discussion (TC=0.645)

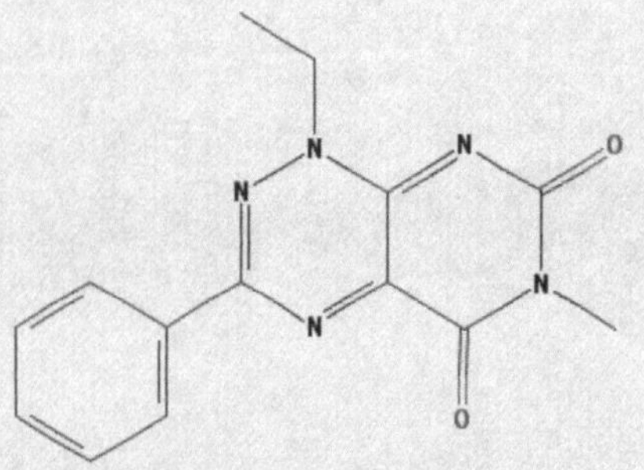

CID 647501(training set-AID 825)
Expt. IC_{50}: 0.27uM

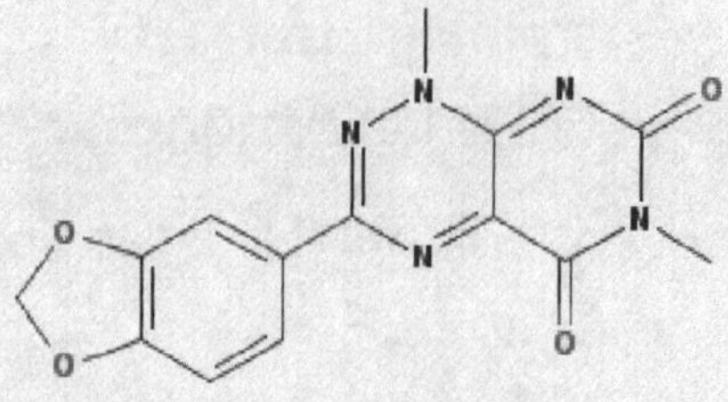

CID 322479(vHTS)
Expt. IC_{50}: 0.08uM
Pred. IC_{50}: 2.11uM

33

Discussion/Conclusion

- Identified possible leads
  - Able to identify non-similar compounds to training set
  - Increased predictive power by retraining models
    - Increased discovery rate
- Models development
  - 27 h on 48 nodes (Intel Xeon E5-2697, 2.7GHz)
- Screening PubChem (72 million compounds)
  - 30 m each screening pass (on 48 nodes)
- Experimental Confirmation
  - 1-2 weeks

34

Acknowledgements

- Choose Ohio First – Bioinformatics
- Dr. Nic Leipzig

- Wednesday 6:00 – 8:00 Poster Session
  - Exhibit Hall 1 -- SLCC

35

433150 Qsars for Predicting Physicochemical and Biochemical Properties of Industrial Chemicals

Abstract:

A current limitation for the introduction of Physiologically Based Bio Kinetic (PBBK) models in the risk assessment arena is the lack of the generic character of these models. In order to expand the applicability domain of PBBK models, model parameterization for data "poor" chemicals is developed using advanced Quantitative Structure-Activity Relationships (QSARs). QSARs are regression or classification models, which form a relationship between the biological effects and the chemistry of a chemical compound and comprise three parts: 1) the activity data to be modeled, 2) the data with which to model and 3) a method to formulate the model. A QSAR model is based on the potency where activity (Y) is a function of one or more descriptors (X) and is used particularly for the estimation of physicochemical properties, biological effects as well as understanding the physicochemical features governing a biological response. The biological effects are normally the property to be modeled, which is linked with the physical or structural chemistry of the molecules.

The methodological approach presented in this study developed on the application of the Linear Free Energy Relationship (LFER), proposed by Abraham, via the incorporation of a large number of chemical compounds, in order to address the biological properties of the main human tissues. This involved the collection of the necessary input data, the statistical analysis and the model implementation to a large number of chemical compounds. The equation was analyzed using two statistical techniques; the Non Linear Regression (NLR) and the Artificial Neural Networks (ANN). Modeling results from the two methods were compared to corresponding literature data for environmental chemicals.

The method of Non Linear Regression was used in order to estimate the parameters in the Abraham's equation. The algorithms selected for fitting the nonlinear regression to the observations was the Least Squares (LS) coupled with the Levenberg-Marquardt algorithm.

Artificial Neural Networks were used for developing a nonlinear model based on Linear Free Energy Relationship. Multi-Layer Perceptron (MLP) model was selected using the scaled conjugate gradient back-propagation algorithm in order to train the network. The multi-layer network consisted of a single input layer, including the values of the molecular descriptors, the experimental values of the physicochemical and the biochemical parameters; one hidden layer and an output layer of a log-sigmoid transfer function. It is noted that the scaled conjugate gradient back-propagation algorithm was selected on the basis of the good convergence rate on the limited sample size used.

The most efficient of the examined methods was Linear Free Energy Relationship analyzed using Artificial Neural Networks. The data set of 33 and 29 industrial chemicals was used in order to predict the tissue/blood partition coefficients for seven main human tissues (heart, adipose, muscle, kidney, lung, liver, brain) and the constants of metabolism (maximal velocity of metabolism, normalized to the human bodyweight and Michaelis – Menten constant), respectively. The initial data sets were divided into the training (70%), the validation (15%) and the test (15%) set. The coefficient of determination, R^2, was equal to 0.97, 0.81, 0.95 for kidney, heart, adipose/blood partition coefficient and 0.92, 0.91, 0.96, 0.81 for liver, muscle, brain and lung/blood partition coefficient, respectively. In addition, the square of correlation coefficient for maximal velocity of metabolism and Michaelis – Menten constant was equal to 0.99 and 0.82, respectively.

The combination of Abraham's equation and ANN method was used for the estimation of tissue/blood partition coefficients for the main human tissues for several chemical compounds with unknown values of partition coefficients in order to expand the chemical space of the application of the developed QSAR model. These compounds were categorized into chemical families, including hydrocarbons, aromatic and halogenated hydrocarbons, alcohols, ketones, ethers and esters, depending on their chemical structure. In order to validate the results obtained from the derived QSAR model, a simpler literature one was used, considering the fractional contents of lipids and water in tissues and blood.

The Linear Free Energy Relationship combined with ANN, provides a generic model, which can estimate a satisfactory number of biological properties, including tissue/blood partition coefficients and metabolic constants. The correlation between the experimental and predicted values obtained using ANN is better than that using NLR. This is because ANN method has the ability to detect and take stock of the non-linear relationships between the independent and dependent variables of the developed model. When significant interactions exist between the input variables, the hidden layer of ANN is used in order to find and analyze them. This cannot easily be accomplished using the NLR method because appropriate transformation of the input or the output parameters is required in order to recognize the nonlinearities and improve model fitness.

The relative importance of input descriptors to the predicted parameters was estimated via the ANN model. It was found that the most critical descriptor for the estimated tissue/blood partition coefficients, except for heart/blood partition coefficient, is the McGowan volume. The distribution of a chemical into a tissue depends on the rate of blood flow to the tissue, the tissue mass, and the partition characteristics between blood and tissue. A compound's distribution between tissues and blood at equilibrium is a function of the respective lipid and water fractions in each matrix (lipid content), which is a function of lipophilicity. Thus, it is a logical conclusion that one of the major determinants of partitioning between blood and tissue is the Mc Gowan volume, as it is a measure of lipophilicity. The polarizability was found to be the most important descriptor influencing the heart/blood partition coefficient. Regarding metabolic parameters, McGowan volume is also the most important parameter for Michaelis – Menten constant, while the most important one for maximal velocity of metabolism is by far the polarizability descriptor. Both Michaelis-Menten constant and maximal velocity of metabolism identify the rate of metabolism of a chemical, which is an enzyme-catalyzed reaction. These reactions are influenced by the ionization state of the substrate of metabolism or the enzyme binding site of the substrate. The maximal velocity illustrates the turnover number of an enzyme, which is the number of substrate molecules converted into product by an enzyme molecule in a unit time when the enzyme is fully saturated with substrate. So, it is directly related to ionization, as well as polarizability.

In general, the results indicated that the molecular descriptors that were inputs of the QSAR model can be suitable for the estimation of the parameters, which characterize the physicochemical and biochemical phenomena. The improved performance of the proposed QSAR model, combining Linear Free Energy Relationshiop and ANN analysis, can be the result of the ability of LFER input parameters to describe the biological responses, as well as the capacity of ANN method to represent mathematically the complex metabolic and physicochemical phenomena.

In general, the proposed QSAR model offers four major advantages: the descriptors of each chemical are easy to obtain, a satisfactory number of biological properties are predicted with the same relationship, it fills the data gaps of data "poor" chemicals, allowing the wide use of PBBK models and it supports the "safe by design" concept for industrial chemicals, by allowing the successful prediction of toxicokinetic behavior based on molecular parameters.

(462h) Statistical Approaches to Crystal Engineering

Adler, P., Southampton University
Coles, S.,
Mapp, L.,
Threlfall, T.,
Woods, D.,

Originally presented on: 11/11/2015 10:45:00 - 11:00:00

Historical publications have left us with a wealth of crystallographic data which dates back for over a century. Further, the collation efforts of the CCDC have permitted the application of big data techniques to the modelling of the behaviour of molecules which may potentially form crystalline systems.

We have been working on using tools championed from applications such as Quantitative Structure Activity Relationships in the sphere of crystalline materials to aid understanding and also to attempt to form predictive models.

Co-crystalline systems are extremely useful in this regard, since the response is easy to test for, and is readily amenable to partition modelling. This will hopefully permit us to tease apart the avalanche of data which has been produced and can be calculated from the experimental data available, and begin to identify which factors are truly important in terms of crystalline material formation.

Such methods are not without their drawbacks, and details will be given on the consequences of entire regions of crystallographic space that have gone unexplored or otherwise unreported- in particular experiments which to most chemists would have yieled unexpected or apparently uninteresting results.

434307 Understanding Protein Transport in Polymer-Grafted Ion Exchangers through Multi-Scale Modeling

Abstract:

When separating proteins using ion-exchange chromatography, matrices with pores with charged polymer grafts have exhibited increased binding capacity, faster adsorption kinetics, and a different diffusion properties compared to traditional macroporous matrices. However, these enhancements depend strongly on the protein properties and concentration as well as the buffer salt concentration. We use multi-scale modeling to study the factors which affect protein transport in polymer-grafted and macroporous materials, which in general depends on both the equilibrium adsorption and diffusion behavior of proteins in the different phases.

We have developed coarse-grained molecular models suitable of proteins diffusing through representative pores of both polymer-grafted and macroporous materials for ion-exchange chromatography. We use a hypothesis-driven, top-down approach, which allows us to access simulations of hundreds of protein and multiple polymer chains out to multiple microsecond time scales.

The kinetic and equilibrium information from these MD simulations is also employed in a simple rate model to study the macroscopic protein concentration profiles in the different materials over time. We apply this multi-scale modeling approach to different types of proteins and adsorbed phases, as understanding how protein transport rates are affected by various protein and material properties should provide insight into the general mechanism by which charged polymers can enhance chromatographic performance.

In simulations of lysozyme interacting with agarose-based matrices with charged dextran polymers, we obtain good agreement with experimental results for equilibrium binding capacities and effective transport rates, as well as their dependence on system parameters such as polymer graft density and surface topology. We observe faster adsorption kinetics and a different adsorption front for the polymer-grafted material as compared to the macroporous material, which is qualitatively consistent with experiments. We also obtain good qualitative agreement with larger proteins such as BSA and IgG, with experimental differences being explainable in terms of molecular details such as volume exclusion and charge distribution. In all cases, the geometry of the surface drastically affects the kinetics of the system, suggesting that idealized smooth surfaces may not be appropriate for modeling surface adsorption in chromatographic materials.

434114 Role of Protein Sequence in Driving Molecular Interactions Between Proteins and Carbon Nanomaterials: A Molecular Dynamics Study

Abstract:

Carbon nanomaterials (CNM's) are commonly used for biological applications, such as drug delivery, biosensors due to their high stability, large surface area and nanoscale dimensions. In order to engineer biologically safer CNMs it is important to understand the molecular interactions of CNMs with proteins which constitute a large part of the biological environment. In our recent combined experimental and molecular dynamics study we found that bovine serum albumin strongly interacts with one of the CNM, graphene nanoribbon (GNR) compared to graphene oxide nanoribbon (GONR). It is often postulated that the driving forces for this association are hydrophobic and van der waals interactions, primarily due to pi-pi stacking of aromatic amino acids of proteins and CNMs. However, there are other factors such as defects and curvature of CNMs, as well as the protein sequence that could have an impact on the protein-CNM interactions. To elucidate the effect of neighboring residues on protein-CNM interactions, we investigate the binding free energy of tripeptides to GNR using biased molecular dynamics simulations. We study X-PHE-GLY tripeptides, where X is a charged/uncharged residue and phenylalanine (PHE) is the chosen reference aromatic residue. In this context, it is expected that the (aromatic) interactions of the GNR with PHE will drive the tripeptide-GNR interactions. By using different residues in the X position we investigate how these driving forces are influenced by protein sequence. Specifically, we have calculated the potential of mean force between the tripeptide and GNR using umbrella sampling. Our results indicate that the strength of binding is influenced by the neighboring residues. Interestingly, the largest free energy of binding was observed for the GLY-PHE-GLY tripeptide. In our talk, we will present detailed results with various tripeptides and discuss the implications of our findings in understanding interactions of proteins with CNMs.

433117 Energy Network Analysis to Quantify Effects of Nanotubes on DNA/Peptide Strength of Interaction

Abstract:

We have developed an energy-based network analysis method, which allows for the characterization of communication pathways in a protein. The method describes protein aminoacids as network nodes, which are connected (or not) based on the average strength of interaction obtained from an equilibrium molecular dynamics simulation. We have shown that this method can accurately describe the ability of a protein to propagate a signal throughout its structure to communicate two domains that are distant from each other. Here, we show that this type of analysis can also be used to quantify the strength of interaction between two molecules, potentially providing a computationally inexpensive alternative to free-energy methods to predict strength of binding. We used our method to measure the effects of nanotube-DNA interaction on the strength of interaction between the same DNA segment and a DNA-binding zinc finger domain peptide. Successful binding of a zinc-finger domain to a DNA segment that is already attached to a carbon nanotube could enable assembly of a system that could function as a biomimetic metabolon for the production of energy in biofuels.

428166 The Role of DNA Sequence in Nucleosome Positioning and Sliding

Abstract:

Nucleosomes provide the basic unit of compaction in eukaryotic genomes and the mechanisms that dictate their position along a DNA sequence are of central importance to epigenetics. In this work, we use molecular models of DNA and proteins to explore the sequence dependent cues that dictate nucleosome positioning. Using high-resolution coarse-grained simulations and free energy calculation techniques, we examine the mechanism by which DNA sequences reposition to find their binding site around the histone protein core. We show that this mechanism is strongly dependent on DNA sequence and that DNA slides around the histone by either a screw-like or loop-like mechanism. These results explain how sequence, manifested as the shape of the DNA molecule, dominates molecular recognition in the problem of nucleosome positioning.

Conformational Thermodynamics of DNA Strands within Electrically Charged Nanopores

Fernando J.A.L. Cruz, José P.B. Mota
LAQV/REQUIMTE, Department of Chemistry, Universidade Nova de Lisboa, 2829-516 Caparica, Portugal
fj.cruz@fct.unl.pt

I. Introduction

Deoxyribonucleic acid (DNA) and single-walled carbon nanotubes (SWCNTs) are prototypical one-dimensional structures, the former in chemical biology and the latter in nanotechnology [1-3]; a plethora of applications currently envisage carbon nanotubes as next-generation encapsulation media for biological polymers, such as proteins and nucleic acids [1]. The interactions between both have been the subject of intense investigation, nonetheless, the corresponding molecular-level phenomena remain rather unexplored. Recently we have shown that, given a sufficiently large hydrophobic nanotube, the confinement of a DNA dodecamer is thermodynamically favourable under physiological environments (134 mM, 310 K, 1 bar), leading to DNA@nanotube hybrids with lower free energy than the unconfined biomolecule [4]. To accommodate itself within the D = 4nm nanopore, DNA's end-to-end length increases from 3.85 nm up to approximately 4.1 nm, via a 0.3 nm elastic expansion of the strand *termini*. The canonical Watson-Crick H-bond network is essentially preserved throughout encapsulation, showing that contact between the DNA dodecamer and the hydrophobic carbon walls results in minor rearrangements of the nucleotides H-bonding. A diameter threshold of 3 nm was established below which encapsulation is inhibited.

It is well known that nanotubes can be electrically charged, either using an AFM tip and applying a voltage bias or by chemically doping the solids with p-type dopants to obtain positively charged solids [5,6]. The effect of charge density upon the energetics and dynamics of confinement needs to be addressed; because DNA's outer surface is negatively charged (phosphate moieties), its interaction with a positively charged solid might lead to the occurrence of encapsulation which is prohibited for purely hydrophobic nanopores. We address this issue using enhanced sampling algorithms (metadynamics, umbrella sampling) to probe the encapsulation mechanism of an atomistically detailed DNA dodecamer (5'-D(*CP*GP*CP*GP*AP*AP*TP*TP*CP*GP*CP*G)-3'), onto positively charged (q = + 0.05 e^{-}/C) SWCNTs of different diameters (3 – 4 nm). In order to allow the extrapolation of results for *in vivo* systems, precise physiological conditions are employed ([NaCl]=134 mM, 310 K, 1 bar).

II. Results & Discussion

In contrast with a purely hydrophobic (40,0) topology (D = 3 nm), the existence of an overall positive charge density on the solid favours the encapsulation of the DNA segment. To probe the thermodynamical stability associated with encapsulation, free-energy landscapes are built using the well-tempered metadynamics scheme [7] and two order parameters relating the distance between centres of mass of DNA and SWCNT, ϕ_1, and the end-to-end length of the biomolecule, ϕ_2. The corresponding Gibbs free-energy maps recorded in Figure 1 show that: *i)* the nanopore endohedral volume ($\phi_1 < 2$) is the thermodynamically preferred region, by comparison with the bulk ($\phi_1 > 2$), *ii)* encapsulated DNA retains its translational mobility, diffusing freely between adjacent free-energy minima located within the solid and *iii)* DNA maintains a *quasi* B-form end-to-end length within the charged topologies (D = 3 – 4 nm). The end-to-end length, Θ, probability distributions, $P(\Theta)$, have been independently probed by umbrella sampling calculations and the results are recorded in Figure 2 for both charged topologies, (40,0) and (51,0), along with the previous results obtained for a purely hydrophobic (51,0) SWCNT [4]. It now becomes clear that charge density on the solid plays a paramount role upon the encapsulation mechanism; the elastic expansion of the double-strand observed for the (51,0) hydrophobic pore (Θ = 4.01 nm) is annihilated when the solid becomes electrically charged, resulting in a maximum probability DNA end-to-end length of Θ = 3.73 nm, consistent with the canonical B-DNA form [8]. The bimodal symmetry associated with the electrostatically charged nanotubes clearly identify the probability maxima corresponding to the equilibrium conformations (Θ = 3.73–3.75 nm), but also two other forms of DNA, at Θ = 3.51 nm (40,0) and Θ = 4.29 nm (51,0), where the former corresponds to a compressed conformation of the double-strand.

A nanoscopic picture of the encapsulated DNA molecule is produced by calculating the corresponding number density maps, as indicated in Figure 3 and obtained from atomically detailed mass histograms (binwidth = 2×10^{-3} nm). Figure 3 reveals the existence of a cylindrical exclusion volume centred along the (51,0) main axis, where molecular density is $\rho \approx 0$, which can be attributed to the strong electrostatic attraction between DNA (PO_4^{3-} ions) and the solid, pulling the former towards the walls and away from the nanopore center. Entropic effects caused by the pore narrowness of the (40,0) SWCNT force the DNA molecule to cluster tightly around the nanopore center, where it exhibits the region of highest molecular density.

As far as we are aware these observations are the first of their kind, and they come to pave the way for the design of smart nanotube based devices for *in vivo* DNA encapsulation.

Acknowledgements. This work makes use of results produced with the support of the Portuguese National Grid Initiative (https://wiki.ncg.ingrid.pt). F.J.A.L. Cruz gratefully acknowledges financial support from FCT/MCTES (Portugal) through grants SFRH/BPD/45064/2008 and EXCL/QEQ-PRS/0308/2012.

References. [1] H. Kumar *et al.*, *Soft Matter* **7** (2011) 5898. [2] B.M. Venkatesan and R. Bashir, *Nature Nano.* **6** (2011) 615. [3] A.D. Franklin *et al.*, *Nano Lett.* **12** (2012) 758. [4] F.J.A.L. Cruz *et al.*, *J. Chem. Phys.* **140** (2014) 225103. [5] X. Zhao and J.K. Johnson, *J. Am. Chem. Soc.* **129** (2007) 10438. [6] F.J.A.L. Cruz *et al.*, *RSC Advances* **4** (2014) 1310. [7] A. Barducci *et al.*, *Phys. Rev. Lett.* **100** (2008) 020603. [8] J.M. Vargason *et al.*, *Proc. Nat. Acad. Sci.* **98** (2001) 7265.

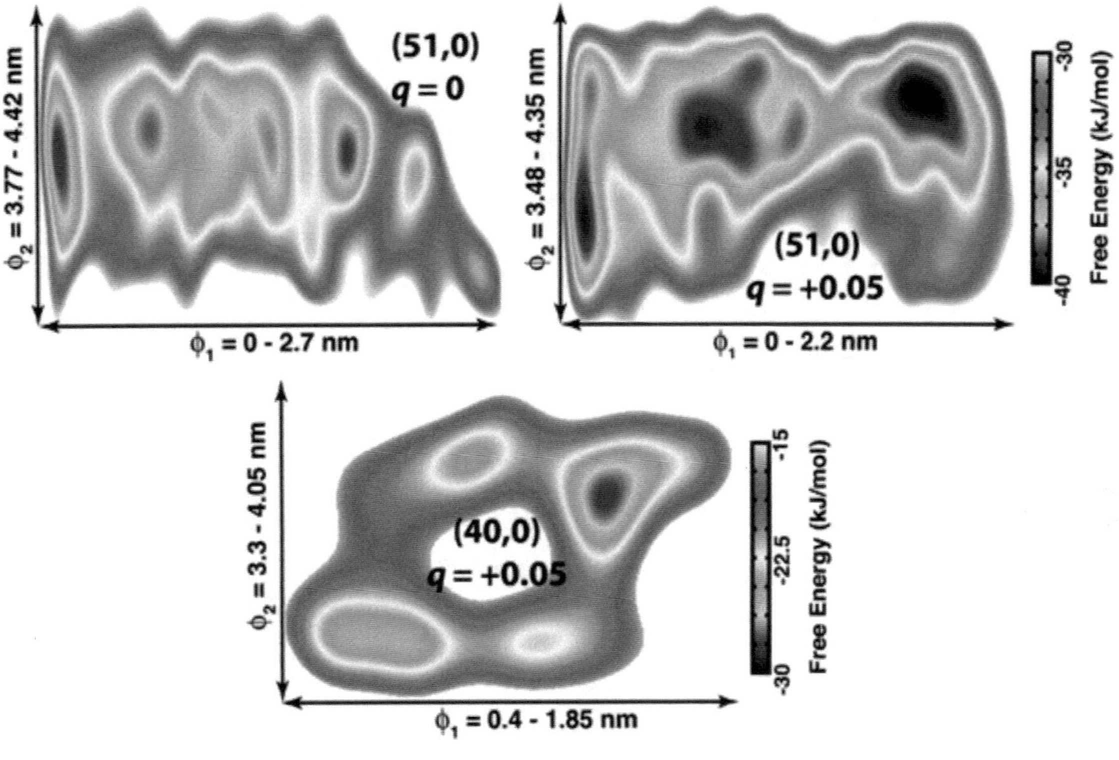

FIGURE 1 – Gibbs free-energy maps of encapsulated DNA. ϕ_1 is the distance between centres of mass of DNA and the SWCNT, projected along the nanopore main axis (z), and ϕ_2 corresponds to the DNA end-to-end length measured between opposite (GC) *termini*. Low-lying free-energy valleys, evidenced as dark blue regions, are always distributed along the nanopore internal volume, $\phi_1 < 2.1$ nm, and the absolute minima are observed at the following (ϕ_1, ϕ_2) pair values: $(0.1, 4.1)_{q=0}^{(51,0)}$, $(1.7, 4)_{q=+0.05}^{(51,0)}$, $(1.4, 3.8)_{q=+0.05}^{(40,0)}$.

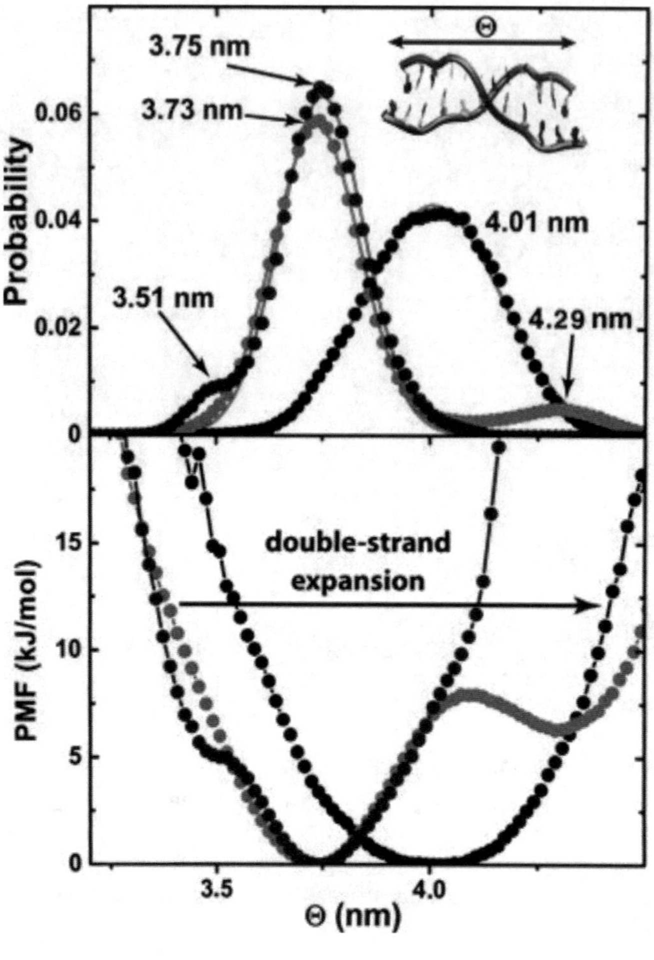

FIGURE 2 – Probability distribution and potential of mean force (PMF) profiles of encapsulated DNA. Because Θ was built as the double-strand end-to-end length, the bimodal symmetry associated with the electrostatically charged nanotubes clearly identify the probability maxima corresponding to the equilibrium conformations (Θ=3.73–3.75 nm), but also two other two other forms of DNA, at Θ = 4.29 nm and a non-canonical conformation at Θ = 3.51 nm. Symbols are the results obtained using umbrella sampling and red lines corresponding to numerical fittings using Gaussian statistics: *black)* (51,0) q = 0, *green)* (51,0) q = +0.05 e^{-}/C and *blue)* (40,0) q = +0.05 e^{-}/C. The Gaussian curve for the (40,0) topology was determined in the range Θ > 3.5 nm.

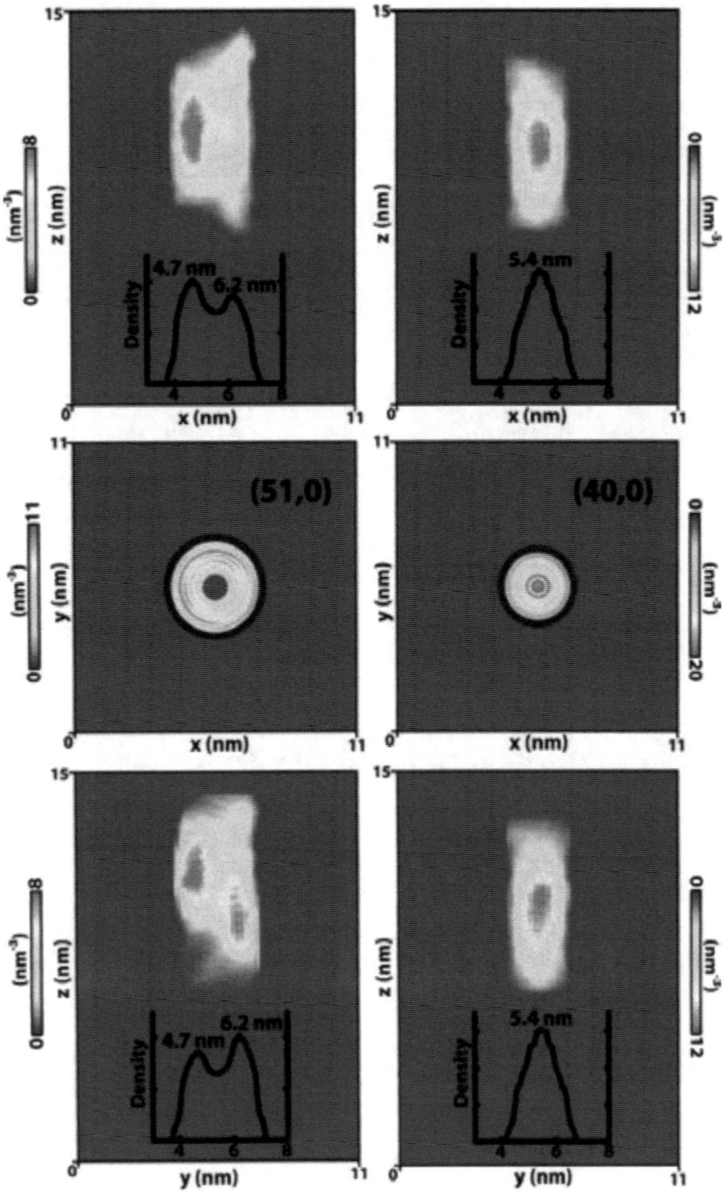

FIGURE 3 – 2D molecular density maps of DNA at the charged topologies. The dashed lines in the middle figures indicate the boundaries of the single-walled carbon nanotubes. The black lines recorded on the inset graphs are 1D density profiles obtained along the corresponding dimension.

Molecular modeling of fibronectin adsorption on topographically nanostructured rutile (110) surfaces

Chun-Ya Wu,[*,a] *Ming-Jun Chen,*[a] *Chuang-Qiang Guo,*[a] *Peter T. Cummings,*[b] *Ting Zheng,*[a,b]

[a] *Harbin Institute of Technology, Harbin, Heilongjiang, China*
[b] *Vanderbilt University, Nashville, TN, United States*

Abstract

To investigate the topographical dependency of protein adsorption, molecular dynamics (MD) simulations were employed to describe the biological behavior of the tenth type-III cell adhesion module of fibronectin (FN–III_{10}) on nanostructured rutile (110) surfaces. The simulation results indicated that the residence time of adsorbed FN–III_{10} molecule largely relied on its binding mode (direct or indirect) with the substrate and the region for protein migration on the periphery (protrusion) or in the interior (cavity or groove) of nanostructures. In the direct binding mode, FN–III_{10} molecules directly connected to the rutile surface were found to be 'trapped' at the anchoring sites, or even penetrate deep into the interior of nanostructures, regardless of the geometrical features presented on substrate. In the indirect binding mode, FN–III_{10} molecules were observed to be indirectly connected to the substrate via a hydrogen–bond network linking FN–III_{10} and interfacial hydrations together. In the indirect mode, our results suggested that a suitable number of facets created by nanostructures that exerted restraints on protein migration played an important role in the stability of the indirect binding between FN–III_{10} and the substrate. However, a doubly unfavorable situation—indirect FN–III_{10}–rutile connections bridged solely by a small number of mediating water molecules and few constraints on movement of protein provided by nanostructures —would result in an early desorption of FN–III_{10} molecules.

Introduction

Adsorption of proteins at a solid–liquid interface is an important phenomenon in biology and highly relevant for the design of implant with excellent biocompatibility. When a biomaterial comes into contact with a biological milieu containing cells, the soluble proteins adsorb onto the surface of biomaterial rapidly, saturating it within a time frame of seconds to minutes.[1] Therefore, living cells may never see the native biomaterial, and the cell–surface interplay is ultimately an interaction between cells and surface–bound protein layer.[2] That is, the cellular response to the surfaces of biomaterials largely depends on the speciation, conformation and orientation of the bound protein layers, whereas the geometrical properties of surfaces can be conveyed indirectly through the adsorbed proteins.[3] Hence, the local changes, especially in the nanoscale surface topographies, may lead to changes in quantity, density and orientation of the adsorbed proteins; thus attention needs to be paid to understand how the protein perceives its substrate.

Many experimental approaches have been applied to figure out whether the existence of nano–scale geometrical characteristics of substrates can enhance protein adsorption, or which kind of nanostructure may act as a preferred site for protein deposition. Kubo et al.[4] created a micropit–and–nanonodule hybrid topography on titanium surfaces, finding out that the nanonodules not only expedited but also boosted the capacity of albumin adsorption. Elter et al.[5] observed that the fibronection tended to adsorb preferentially in the concave grooves on the surface of boron-doped silicon wafers. Chen et al.[6]

found that the adsorbed fibrinogen regularly distributed most on the flanks and valleys of the dot–like protrusions, but sprinkled throughout the flat surface of poly(dimethylsiloxane). The current researches suggest that proteins are sensitive to the nanoscale environment of anchoring surface, probably even showing special affinity for certain type of nanostructures.

Although the protein-surface systems cover a huge diversity, a detailed understanding of how the individual proteins attach to artificially nanostructured surfaces on a molecular level is urgently required. Compared to the experimental approaches, computational methods, which could eliminate occasional influences of experiments and create plausible structural models of proteins on solid surfaces, seems to be a feasible option to reveal the interaction mechanism between the surface of biomaterials and proteins. Monti[7] and Walsh[8] conducted molecular dynamic (MD) simulations to understand the adsorption of polypeptides on the strongly polar surfaces with explicit water, observing that binding is not to the surface itself, but rather to a structured interfacial water layer. Yang et al.[9] believed that the main driving force for protein binding on the titania nanotubes derived from the physical adsorption on the basis of MD simulations. Our group performed MD simulations to figure out the influence of cation mediation in peptide-surface interaction, observing that divalent cations seemed more effective than monovalent cations in bridging the negatively charged adsorbate to the negatively charged adsorbent.[10,11]

In the present work, MD simulations were also utilized to describe the early adsorption of a protein fragment onto the nanostructured surfaces. The 10th Type III module of fibronectin (FN-III$_{10}$), which contains the important cell–binding sequence—arginine–glycine–aspartic acid (Arg–Gly–Asp, RGD), was selected as the adsorbate,[12] and the most stable phase of TiO_2 polymorphs at ambient temperature—rutile[13] was chosen to be the adsorbent. As is well known, the hydroxyl coverage of a native oxide film formed on a Ti-surface depends on the adsorption state of water; however, the fundamental question that whether water dissociates upon adsorption on the TiO_2 surface remains controversial. The common view in experimental researches[14,15] is that water adsorbs molecularly on the ideal TiO_2 surface and dissociation does only take place at defect sites, whereas a delicate balance between dissociated, partial dissociated, and molecular adsorption of water is suggested in theoretical studies.[16,17] In light of this, it is hard to gain a clear-cut answer to whether water adsorbs associatively or there is dissociation of water to produce hydroxyl sites on the TiO_2 surface. Moreover, the focus of this work is to assess the influence of geometrical features of substrate on the adsorption behavior of protein, thus three nonhydroxylated rutile (110) surfaces characterized with different types of nanostructures (i.e., protrusion, cavity or groove) were introduced here.

Simulation methods

The MD simulations of FN-III$_{10}$–rutile assembly in aqueous solution were carried out in the NVT ensemble. The rutile (110) surface was reproduced through the Matsui and Akaogi parametrization,[18] and the details about the size of a single protrusion, cavity and groove are shown in Figure 1. The Amber force field[19] was utilized to describe the structure of protein, and the SPC/E water model[20] was chosen to represent the water molecules. The interaction potentials between amino acid residues/water and rutile atoms were obtained from the literature,[21] and the parameters of Lennard-Jones potential for the cross interactions between other nonbonded atoms followed the Lorentz-Berthelot rule.[22] The simulation box with the size roughly equal to 142×130×80 Å^3 was filled with water molecules, the density of which was set to the value corresponding to 1 atm of solution pressure. Periodic boundary conditions were applied in the *x*– and *y*–direction, and a reflecting boundary condition was applied in the *z*–direction. The Nosé–Hoover thermostat[23] was used to control the temperature and the particle–particle particle–mesh (PPPM) solver[24] was employed to handle the long–rang electrostatic interactions.

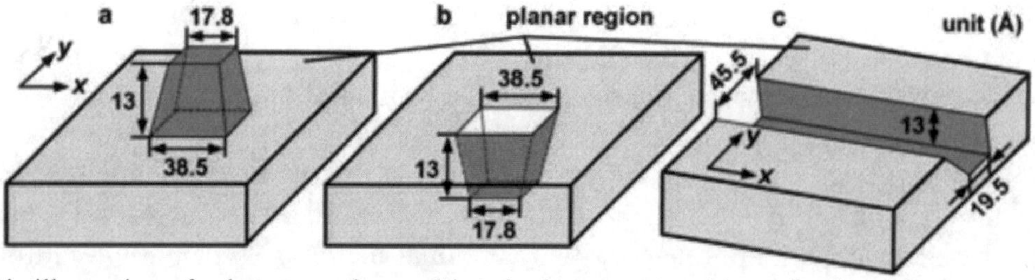

Fig. 1 Schematic illustration of substrate surfaces with a single protrusion (a), cavity (b) and groove (c).

The FN-III$_{10}$ molecules were put close to the nanostructured rutile surfaces with different arrangement, maintaining the initial distances from any atom of FN-III$_{10}$ to the rutile atoms larger than 4.5 Å. At the beginning of MD simulations, the solute (FN-III$_{10}$ and rutile) was kept frozen to randomize the position of water molecules, but the constraints were removed after an energy minimization (corresponding to approximate 0 K), only keeping the lowest layer of rutile fixed till the end of simulations. The system temperature was then increased gradually from 0 K to 300 K in increments of 50 K every 250 ps, with the FN-III$_{10}$ molecule treated as an independent rigid body, which moved and rotated as a single entity (i.e., body movement). After that, the systems were relaxed at T=310 K for 1 ns with position restraints on the backbone atoms of amino acid residues (i.e., restraining movement), and the resulting configurations were equilibrated without any constraint for 8~16 ns, depending on the specific state of different molecular assemblies.

Results

In the MD simulations, FN-III$_{10}$ molecules were initially put close to the rutile surface without direct connection. The proteins remained in the immediate vicinity of the presented nanostrucures after the 'body' and 'restraining' movements were the focus of this work, and seven typical configurations shown in Figure 2 were selected to proceed to the production run. Proteins can 'sense' the topography of surfaces at the nanometer scale;[25–28] but the research to gain a detailed understanding of how different surface nanostructures affect the binding affinity of proteins is still in its infancy. Therefore, the emphasis here was to test whether the protein can keep within the region of nanostrucures, or even break through the screen of interfacial water layers, forming direct interactions with the rutile surface.

Adsorption behavior of FN-III$_{10}$ molecule on protrusions

Three FN-III$_{10}$–protrusion assemblies were introduced with the protein located above the planar region, at the juncture of flank and planar region, and on the top of summit, respectively. For clarity, the three FN-III$_{10}$ molecules were called 'FN-III$_{10}$-P1','FN-III$_{10}$-P2', and 'FN-III$_{10}$-P3' in sequence. The evolution of minimum distances from FN-III$_{10}$ to rutile surface are shown in Figure 3. FN-III$_{10}$-P1 remained in the vicinity of both flank and planar region at an early stage of production run, moving toward the bulk solvent after t=4 ns. FN-III$_{10}$-P2 stayed within the planar region for another 1.5 ns after its departure from the flank of protrusion. The minimum distances from FN-III$_{10}$-P3 to the summit and flank fluctuate widely during the early stage of equilibration period, but reach a plateau after t=5 ns. The minimum FN-III$_{10}$–summit distance of ~2.0 Å indicates that direct connections between FN-III$_{10}$-P3 and the summit of protrusion should have been formed during the final 3 ns of MD simulation.

From the trajectory, it can be seen that FN-III$_{10}$-P1 presented a 'standing up' configuration, keeping the C-terminal carboxyl group hydrogen–bonded to the first– and second–layer water molecules during the final 4 ns of equilibration stage. However, the indirect FN-III$_{10}$–rutile connections mediated by very few water molecules were too weak to 'trap' the protein on the surface. On the contrary,

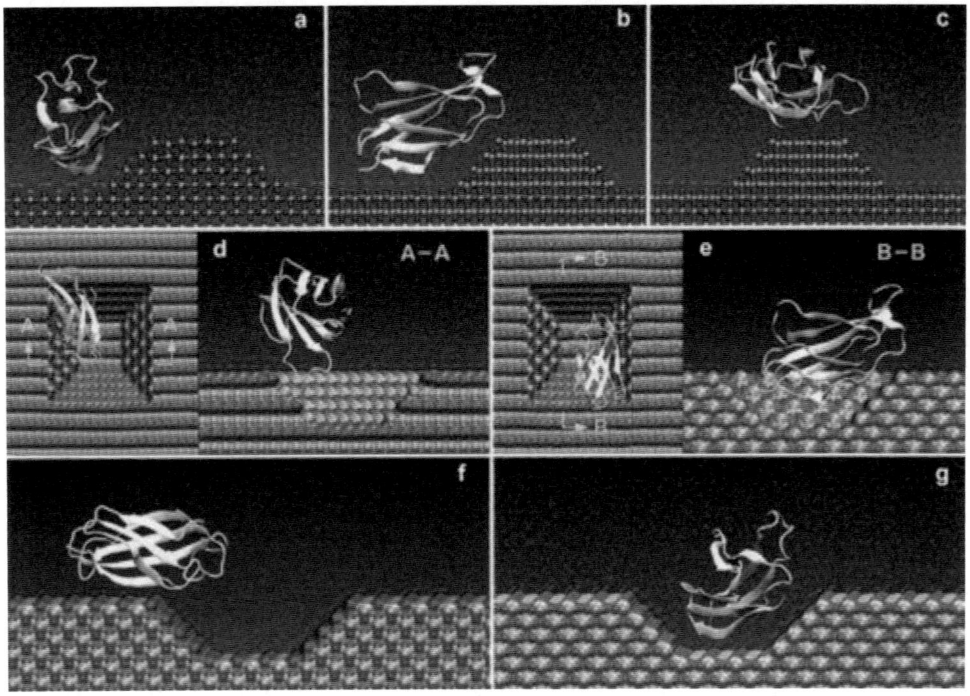

Fig. 2 Initial configurations of the FN-III_{10}-rutile assemblies. FN-III_{10} on the rutile surface with a protrusion (a)~(c), with a cavity (d) and (e), with a groove (f) and (g). FN-III_{10} is displayed in NewCartoon mode of VMD,[29] TiO_2 atoms in (a)~(c) are displayed in VMD's CPK mode, TiO_2 atoms in (d)~(g) are displayed in VMD's surf mode. Left snapshots in (d) and (e) are top views, right snapshots in (d) and (e) are sectional views.

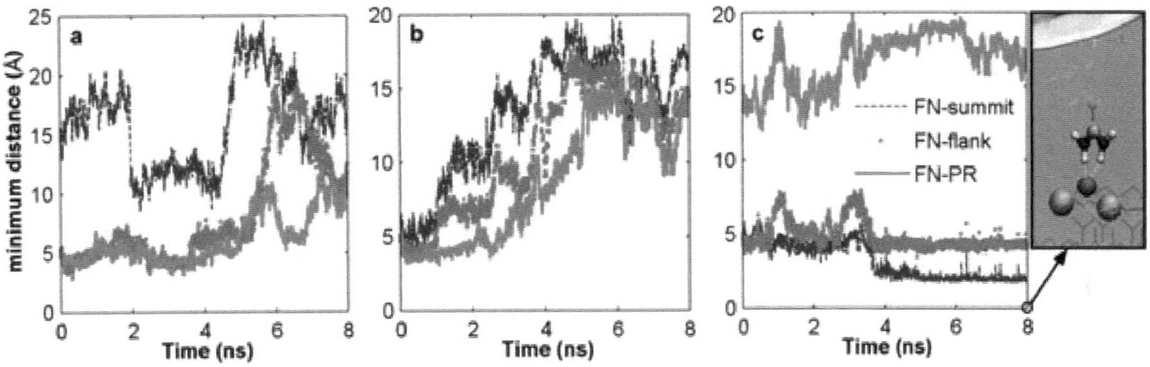

Fig. 3 Evolution of minimum distances from FN-III_{10}-P1 (a), FN-III_{10}-P2 (b), and FN-III_{10}-P3 (c) to different parts of rutile surfaces with a single protrusion. PR stands for planar region.

FN-III_{10}-P2 exhibited a 'lying down' configuration, possessing a relatively larger contact area with the planar region; thus the protein could remain there for a longer time after its departure from the flank. The FN-III_{10}-P3 finally formed direct hydrogen bonds with the doubly coordinated bridging O atoms (O_{BO}) via the guanidine group in residue Arg93. The guanidine group would occasionally hop along the [001] direction, but the H_{NH_2}–O_{BO} distance basically met the criterion of hydrogen bond.

Adsorption behavior of FN-III_{10} molecule on cavities

For the rutile surface with a single cavity, the 'FN-III_{10}-C1' located at the juncture of inwall and planar region, and the 'FN-III_{10}-C2' settled inside the cavity were introduced, respectively. The minimum distance from FN-III_{10}-C1 to the valley of cavity (Figure 4) indicates that the protein remained more than 9 Å away from the valley within the first 3.7 ns, but the former continued to approach the latter till the minimum distance between them plateaued out after t=6.5 ns. The snapshots in Figure 4 show that FN-III_{10}-C1 began to bind with the left inwall rather early. The entire protein flipped over around the anchoring site, finally reaching the valley to form hydrogen bonds with the interfacial water molecules; thus, the anchoring site is of vital importance in determining the subsequent motion of adsorbate. To preferentially monitor the dynamic events occurring at the anchoring site, Figure 5 displays the evolution of distances from the carbonyl and amide groups in the side chain of residue Asn42 to the top–layer atoms of cavity inwall. It can be seen that a water molecule was initially bound to a 4–coordinated Ti atom of inwall (Ti_1), intermediated between the carbonyl oxygen (O_{CO}) and the Ti_1 atom (left inset), whereas the amide group located far away from the oxygens of inwall (O_1 and O_2). At t=0.12 ns, the water molecule adsorbed on Ti_1 atom reoriented to make room for the Ti_1–O_{CO} bond, which subsequently acted as a fulcrum for the other parts of protein. The amide group of ASN42 swung mildly around the Ti_1–O_{CO} bond until the distance from it to O_1 or O_2 atom was close enough to form hydrogen–bonded connections (right inset).

FN-III_{10}-C2 was located within the cavity in the primary stage of production run, but Figure 6a shows that the FN–valley minimum distance remained generally larger than 5 Å, whereas the FN–inwall minimum distance fell sharply to 2.3 Å at t=5.0 ns. Therefore, it can be inferred that although the protein was not able to form direct interactions with the top–layer surface atoms or the first–hydration shell on the valley, it may be directly bonded or hydrogen-bonded to the surface atoms on the inwall. Figure 6b shows the radial distribution function (RDF) curves calculated for the specific protein atoms and the top-layer TiO_2 atoms of inwall (Unless otherwise stated, the RDF data in this work were averaged from the final 2 ns results of production run). The O_{si}–N_{NH3^+} RDF exhibits a first sharp peak centering at ~2.95 Å, meanwhile the magnitude of Ti_{si}–O_{COO^-} RDF begins to increase when r exceeds 4.0 Å. That is, at least one NH_3^+ group was located within the immediate proximity to the O_{si} atoms, allowing the formation of N_{NH3^+}...O_{si} hydrogen bonds, and the trajectory confirms that the NH_3^+ group in residue Arg6 did break through the water barriers, forming hydrogen bonds with O_{si} atoms on the inwall. However, the carboxyl groups were not capable of forming direct bonds with the Ti_{si} atoms, but five carboxyl groups (the involved residues include Glu9, Asp23, Glu38, Glu95 and Ile96) remained in the interior of cavity, forming hydrogen bonds with the first–hydration shells of inwalls. Meanwhile, the carboxyl group of Glu95 was stably bridged to the valley of cavity via the second–layer water molecules, thus an intensive hydrogen-bond network linking the FN-III_{10}-C2 and the interior of cavity was formed.

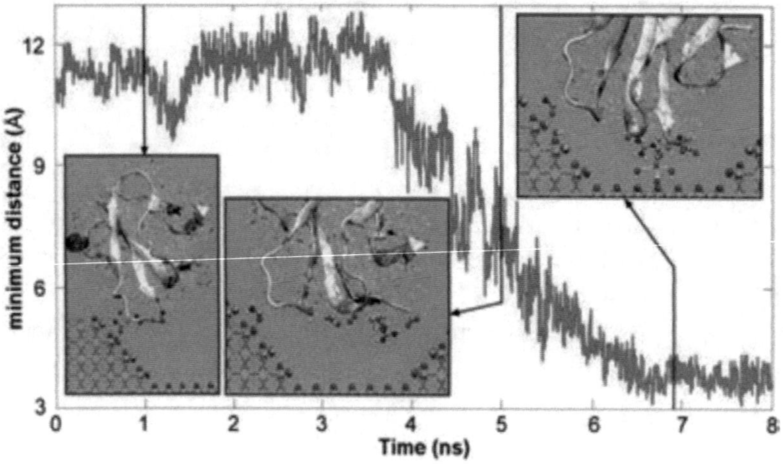

Fig. 4 Evolution of minimum distance from FN-III_{10}-C1 to the valley of cavity, together with protein conformations at different time points. To clearly display the interior of cavity, parts of inwalls are hidden in the snapshots.

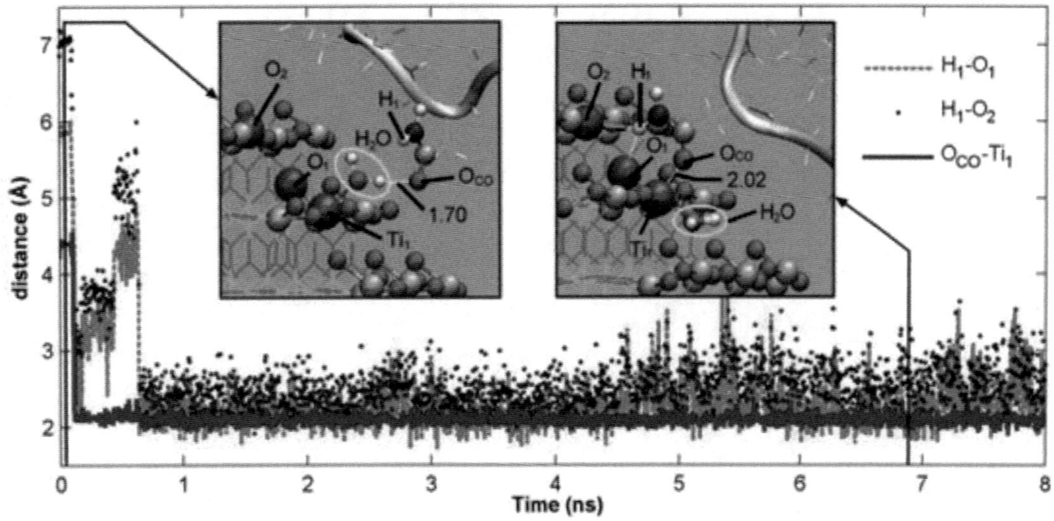

Fig. 5 Evolution of distances from carbonyl and amide groups in the side chain of Asn42 (FN-III$_{10}$-C1) to the specified top-layer TiO_2 atoms on the inwall. To clearly display the interior of cavity, parts of inwalls are hidden in the snapshots.

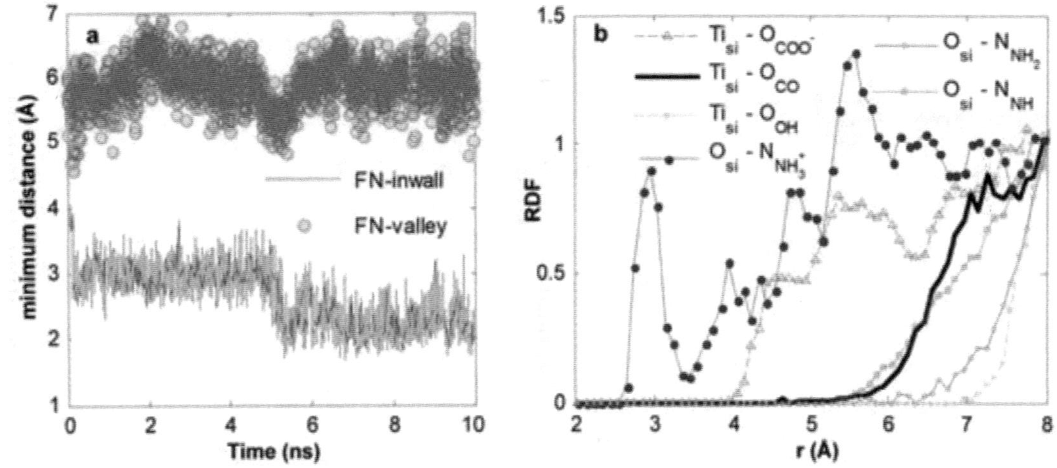

Fig. 6 Evolution of minimum distances from FN-III$_{10}$-C2 to the inwall, valley of cavity (a) and RDFs between specific protein atoms and TiO_2 atoms (b). Ti_{si} and O_{si} refer to the top–layer Ti and O on the inwall of cavity, respectively.

Adsorption behavior of FN-III$_{10}$ molecule on grooves

The grooves run through in the *x*–direction with the width larger than that of cavities, resulting in a more extensive territory for protein activity. It can be supposed that the probability of FN-III$_{10}$ arrive in the valley of groove is greater than the case in the cavity, since the proteins surrounded by four inwalls of cavity could readily interact with the constraint facets, which may hinder them moving further toward the valley. Therefore, two FN-III$_{10}$-groove assemblies were selected, including one adsorbate located at the juncture of inwall and planar region, and the other stayed on the top of groove valley. For clarity, the two FN-III$_{10}$ molecules were called 'FN-III$_{10}$-G1' and 'FN-III$_{10}$-G2', respectively.

The minimum distances from FN-III$_{10}$-G1 to the grooved surface, and the typical conformations are shown in Figure 7. The protein was originally far away from the valley, lying at the juncture of inwall and planar region. The NH_3^+ in the side chain of residue Lys86 was the only functional group interacting with the substrate, via the mediation of the first– and second–hydration shells on the left inwall of the groove. FN-III$_{10}$-G1 gradually detached from the rutile surface after 3 ns, but returned at t=~11 ns with the height of protein varying slightly during the later stage of production run. However, the link between the entire protein and the grooved surface was extremely weak, only depending on a few mediated water molecules. For FN-III$_{10}$-G2 contained in the interior of groove, RDFs between four carboxyls in the interior of groove, all the amino groups and first–hydration shells were calculated to determine the binding modes of FN-III$_{10}$–rutile assembly (Figure 8). Referred from the positions of first–peak centers, COO^-_3 should be the only functional group forming hydrogen bonds with the first hydration on the valley, whereas COO^-_1, COO^-_2 and COO^-_3 could be hydrogen bonded to the first hydration on the inwall without exception. However, the magnitudes of first peaks indicate that the persistence of hydrogen bonds is in the following order: $COO^-_3...O_w>COO^-_2...O_w>COO^-_1...O_w$ (O_w refers to oxygens of the first–hydration waters). Moreover, amino groups could occasionally get within close proximity to the first–hydration shell on the inwall, but stable hydrogen bonds were never formed. To visualize the FN-III$_{10}$–rutile interaction, Figure 9 gives the typical conformations of FN-III$_{10}$-G2. The entire protein lay sprawled across the valley at t=13 ns, interacting with both inwalls at the same time. Plenty of waters in the surface hydration shells were involved in the hydrogen–bond network with FN-III$_{10}$-G2. However, the protein stayed farther away from the right inwall at t=15 ns, leaving only the COO^-_3 connected to the right inwall and valley of groove indirectly via the mediation of first– and second–layer water molecules, which agrees well with the order of persistence of $COO^-...O_w$ hydrogen bonds.

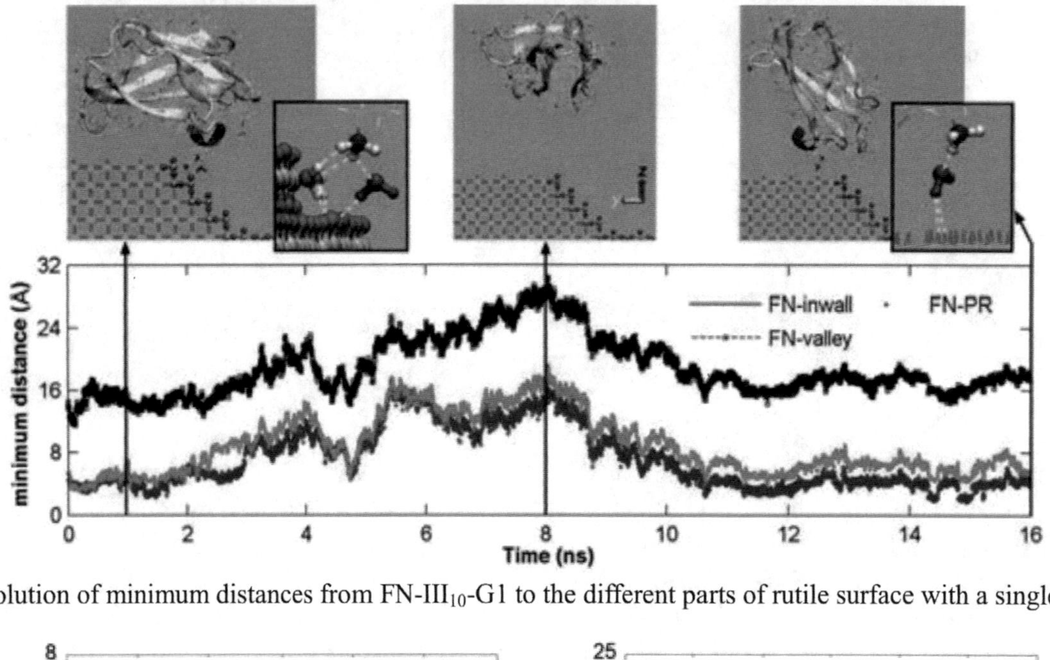

Fig. 7 Evolution of minimum distances from FN-III$_{10}$-G1 to the different parts of rutile surface with a single groove.

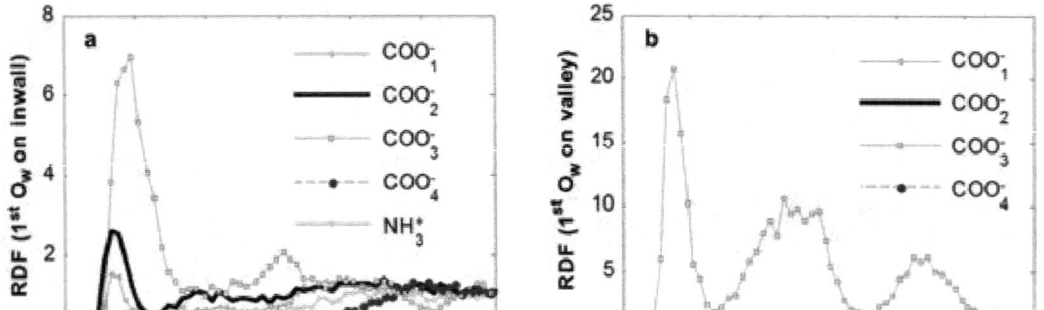

Fig. 8 RDFs between functional groups of FN-III$_{10}$-G2 and the first–layer O_w on inwall (a) and valley (b) of groove. COO^-_1~COO^-_4 refer to the carboxyl groups of residue Glu9, Asp23, Glu95 and Ile96 in sequence.

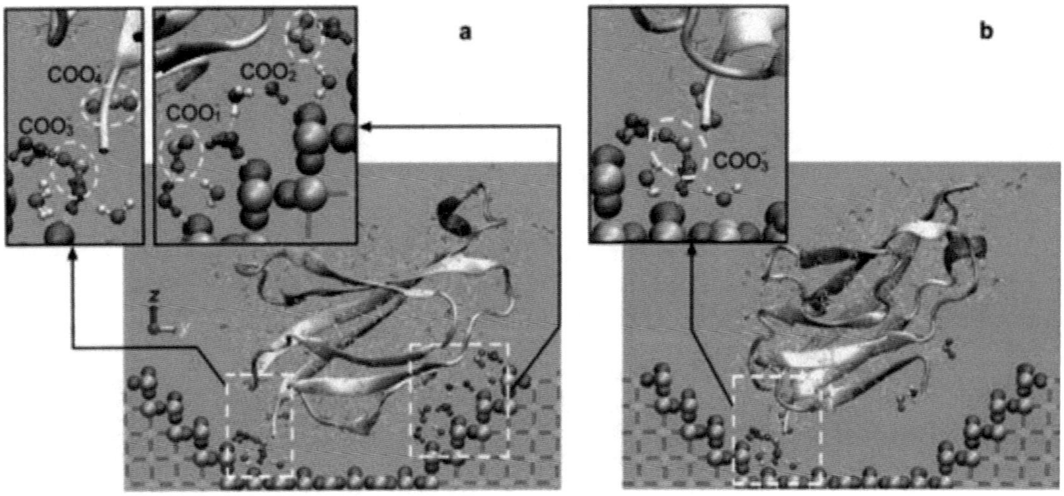

Fig. 9 Conformations of FN-III$_{10}$-G2 on the grooved surface at 13.0 ns (a) and 15.0 ns (b). FN-III$_{10}$ is displayed in combined mode of VMD (NewCartoon and line mode), functional groups in close proximity to substrate and first–/second–layer water molecules directly interacting with FN-III$_{10}$ are displayed in VMD's CPK mode, the TiO_2 atoms locating at the edge of groove and the residual ones are displayed in VMD's VDW and line mode, respectively. β-strand, α-helix, 3_{10}-helix and second–layer mediating water molecules are highlighted by yellow, violet red, dark blue, and purple, respectively. The residual water molecules are omitted for clarity.

Discussions

Lord et al.[3] stated that the detailed shape of nanofeatures was of paramount importance to protein adsorption, in addition to the spatial distribution. As they documented, the adsorption stability of protein was found to rely on the geometrical features of nanostructures to a great extent in this work. The protein segment in close proximity to protrusion generally interacts with up to one flank. Combined with the summit or planar region, there are two facets from the substrate capable of imposing position constraints on the protein at the utmost. The groove runs through in the *x*–direction and hence a total of three constraint facets, i.e., two inwalls plus valley, will contribute to the limited migration of protein. By contrast, the number of facets imposing restraints on the adsorbed protein reaches five for the cavity, including four inwalls and one valley. Therefore, the territory size for protein migration on the periphery (e.g., protrusion) or in the interior (e.g., cavity or groove) of nanostructures can be ordered as follows: protrusion> groove>cavity, which may explain the experimental results gained by Elter et al.[5] that concave grooves seemed more favorable for fibronectin adsorption than convex summits. The convex summits in Elter's experiments resemble the presented protrusion in overall shape, and the concave grooves designed by Elter et al. are similar to the state of the created grooves.

FN-III$_{10}$-P1 and FN-III$_{10}$-P2 could remain hydrogen–bonded to the first– and second–hydration shells of the protrusions during the early stage of production runs, but both of them detached from the substrate gradually. This is because although the positions of first–hydration waters were fixed due to strong attractions from surface Ti atoms, the number of them engaged in mediation of protein adsorption was extremely small. As for the second–hydration shell, the mobility of waters will result in weak hydrogen–bonded connections with the protein even if an abundance of second–hydration waters are involved, not to mention only a small handful of mediating water molecules available here. That is, when the protein is in a doubly unfavorable situation, e.g., bridged to the rutile surface indirectly by a minority of water molecules and under insufficient position constraints from the nanostructures, it is hard to be remained in the proximity of substrate.

Unlike the above case, when direct interactions can be formed between the protein and the substrates, the results will turn out to be completely different. Take FN-III_{10}-C1 on the rutile surface with a cavity, for instance. The coordination of Ti_{l} on the left inwall did not approach saturation even with one adsorbed water molecule; thus the O_{CO} atom was able to bind with Ti_{l} before other waters could penetrate into the interspace between them. Consequently, the Ti_{l}–O_{CO} bond acted as an anchoring site, around which the entire protein flipped over towards the valley of cavity to actively interact with the interfacial hydrations. The same goes for FN-III_{10}-P3 on the protrusion and FN-III_{10}-C2 in the cavity in a different form. These two proteins broke through the water barriers, forming hydrogen bonds with O_{BO} on the summit edge and O_{si} on the inwall, respectively. The long–term nature of protein adsorption could not be unequivocally determined from the short, sub–20ns simulations reported here; nevertheless the hydrogen–bonded connections remained stable over the time scale of the present simulations. Therefore, it can be inferred that direct interplays between FN-III_{10} and rutile are capable of 'trapping' the protein on the surface of substrate, or even promoting it to penetrate deep into the interior of nanostructures, no matter how many constraint facets will be provided to limit the mobility of adsorbed protein. In addition, another favorable condition for protein adsorption is that an intensive hydrogen–bond network could be developed to effectively link the FN-III_{10} and the interfacial hydrations together, while a suitable number of constraint facets are offered—FN-III_{10}-G2 in the groove for example. The protein rotated and moved inside the groove, inevitably approaching one side of the inwall when leaving the other, as a result of position restraints from both sides. Ultimately, FN-III_{10}-G2 was confined to the interior of groove, remaining bridged to the substrate surface by the first– and second–hydration mediating waters.

Conclusions

MD simulations were employed in the present study to gain a detailed understanding of the interaction between a fibronectin segment (FN–III_{10}) and the artificially nanostructured rutile (110) surfaces. Protrusions, cavities and grooves were introduced to the surface models to emulate the nanoscale environment of anchoring substrates. The binding modes of FN–III_{10}–rutile assembly were of paramount importance in affecting the stability of protein adsorption. Once the protein formed direct connections with the surface titaniums or broke through the water barriers to form hydrogen-bonds with the surface oxygens, it would tend to stay in close proximity to the substrate, or even penetrate deep into the interior of nanostructures. Moreover, the detailed shape of the nanostructures, which determines how many constraints could be imposed on the adsorbates by facets, was also demonstrated to strongly influence the residence time of adsorbed proteins within the nanostructured regions. In the case that the FN–III_{10} could only be bridged to the substrate indirectly via interfacial hydrations, the situation may also be favorable for protein adsorption on the premise that abundant water molecules are involved in mediation and sufficient constraining facets are available to limit the mobility of adsorbed protein. On the contrary, sparse hydrogen–bonded connections between proteins and interfacial hydrations, combined with inadequate constraints from the nanostructures, would lead to a relatively large region for protein migration, facilitating the desorption of FN–III_{10} molecules.

The findings presented in this work provide important insight into the correlation between protein adsorption and nanostructured topography of surface on a molecular level. Although it is still impossible to predict which type of nanoscale geometrical features will be most effective in promoting/inhibiting protein adsorption, thereby eliciting a particular cellular response, the introduction of tailored nano-topographies in biomedical products seems to be a potentially promising approach to improve their performance. In addition, the synergistic effect between the chemistry, topography and mechanical properties of biomaterial surfaces should be taken into account when seeking an effective control of cellular behaviors.

References

1 R. A. Latour, *Biointerphases*, 2008, **3**, FC2–FC12.
2 B. Kasemo, *Surf. Sci.*, 2002, **500**, 656–677.
3 M. S. Lord, M. Foss and F. Besenbacher, *Nano Today*, 2010, **5**, 66–78.
4 K. Kubo, N. Tsukimura, F. Iwasa, T. Ueno, L. Saruwatari, H. Aita, W. A. Chiou and T. Ogawa, *Biomaterials*, 2009, **30**, 5319–5329.
5 P. Elter, R. Lange and U. Beck, *Colloid Surf. B-Biointerfaces*, 2012, **89**, 139–146.
6 H. Chen, W. Song, F. Zhou, Z. Wu, H. Huang, J. H. Zhang, Q. Lin and B. Yang, *Colloid Surf. B-Biointerfaces*, 2009, **71**, 275–281.
7 S. Monti, *J. Phys. Chem. C*, 2007, **111**, 16962–16973.
8 A. A. Skelton, T. Liang and T. R. Walsh, *ACS Appl. Mater. Interfaces*, 2009, **1**, 1482–1491.
9 W. Yang, X. Xi, X. Shen, P. Liu, Y. Hu and K. Cai, *J. Biomed. Mater. Res. Part A*, 2014, **102A**, 3598–3608.
10 C. Y. Wu, A. A. Skelton, M. J. Chen, L. Vlček and P. T. Cummings, *Langmuir*, 2012, **28**, 2799–2811.
11 C. Y. Wu, A. A. Skelton, M. J. Chen, L. Vlček and P. T. Cummings, *J. Phys. Chem. C*, 2011, **115**, 22375–22386.
12 M. D. Pierschbacher and E. Ruoslahti, *Nature*, 1984, **309**, 30–33.
13 W. H. Baur and A. A. Khan, *Acta Cryst.*, 1971, **B27**, 2133–2139.
14 F. Allegretti, S. O'Brien, M. Polcik, D. I. Sayago and D. P. Woodruff, *Phys. Rev. Lett.*, 2005, **95**, Article ID: 226104.
15 D. Brinkley, M. Dietrich, T. Engel, P. Farrall, G. Gantner, A. Schafer and A. Szuchmacher, *Surf. Sci.*, 1998, **395**, 292–306.
16 P. J. D. Lindan and C. Zhang, *Phys. Rev. B*, 2005, **72**, Article ID: 075439.
17 W. Zhang, J. Yang, Y. Luo, S. Monti and V. Carravetta, *J. Chem. Phys.*, 2008, **129**, Article ID: 064703.
18 M. Matsui and M. Akaogi, *Mol. Simul.*, 1991, **6**, 239–244.
19 W. D. Cornell, P. Cieplak, C. I. Bayly, I. R. Gould, K. M. Merz, D. M. Ferguson, D. C. Spellmeyer, T. Fox, J. W. Caldwell and P. A. Kollman, *J. Am. Chem. Soc.*, 1995, **117**, 5179–5197.
20 H. J. C. Berendsen, J. R. Grigera and T. P. Straatsma, *J. Phys. Chem.*, 1987, **91**, 6269–6271.
21 V. Carravetta and S. Monti, *J. Phys. Chem. B*, 2006, **110**, 6160-6169.
22 M. P. Allen and D. J. Tildesley, In Computer Simulation of Liquids, 1st ed.; Oxford University Press Inc.: New York, 1989; p 22.
23 W. G. Hoover, *Phys. Rev. A*, 1985, **31**, 1695–1697.
24 R. W. Hockney and J. W.Eastwood, In Computer Simulation Using Particles, 1st ed.; Taylor & Francis Group: New York, 1988; p 22.
25 X. Jiang, L. Shang, Y. L. Wang and S. J. Dong, *Biomacromolecules*, 2005, **6**, 3030–3036.
26 P. Roach, D. Farrar and C. C. Perry, *J. Am. Chem. Soc.*, 2006, **128**, 3939–3945.
27 A. A. Vertegel, R. W. Siegel and J. S. Dordick, *Langmuir*, 2004, **20**, 6800–6807.
28 D. P. Song, M. J. Chen, Y. C. Liang, Q. S. Bai, J. X. Chen and X. F. Zheng, *Acta. Biomater.*, 2010, **6**, 684–694.
29 W. Humphrey, A. Dalke and K. Schulten, *J. Mol. Graph. Model.*, 1996, **14**, 33–38.

422689 Theoretical Insight into the Behavior of Bacteriocins in a Model Membrane Environment

Abstract:

Antibiotic resistance has emerged as a serious threat in global human health over the last years. Various studies estimate that if no action is taken, we will be soon entering an era when even the most insignificant bacterial infections could lead to prolonged hospitalization or even death. Therefore, there has been an increased effort to discover new antibacterial therapies. One alternative to conventional antibiotics is the use of antimicrobial peptides (AMPs). Various organisms, such as bacteria, plants and humans, produce AMPs as part of their defense mechanism to ward off bacteria. In order to exploit the full potential of AMPs and establish them as new antibiotic agents, a detailed understanding of their bactericidal mechanism of action would be useful. Molecular simulations have been extensively used, to provide a biophysical insight into the behavior of AMPs toward unravelling their complicated mechanism of action.

We focus on AMPs that are produced by bacteria. These are commonly known as bacterocins and have been shown to be effective in the treatment of relevant bacterial infections. We have examined the interactions of Enteroccin A, a class IIa bacteriocin, and Plantaricin EF, a dimeric class IIb bacteriocin, on the presence of a model lipid bilayer that mimics bacterial cell membranes. We have employed atomistic molecular dynamics simulations and enhanced sampling techniques in order to investigate how each of these peptides interacts with the membrane, when they are placed under different initial configurations, such as on the surface or inside the membrane. We finally have simulated mutants of the wild type Plantaricin EF and observed a similar trend between experimental activity and biophysical behavior.

Our simulation results indicate that there are specific groups of residues and motifs that govern the biophysical behavior of the bacteriocins examined. The most important motif is the GxxxG motif, which is also shared among many types of bacteriocins and is hypothesized to be connected with the ability of peptide dimers to penetrate the membrane. We are currently studying the influence of the GxxxG motif on the dimerization process of bacteriocins employing atomistic molecular dynamics simulations and potential of mean force calculations.

We believe that the information yielded from these simulations can provide valuable information for the design and optimization of new antibiotic agents founded on bacteriocins.

424063 Unmasking the Mystery Base Employed By the T. Reesei Cel6A Cellulase

Abstract:

Trichoderma reesei Cel6A is an industrially-important enzyme for converting cellulose to renewable fuels and chemicals. Still, many questions remain for the understanding of its catalytic mechanism despite many structural and kinetic studies with both wild-type and mutant enzymes. Cel6A is a cellobiohydrolase (CBH) that binds the cellulose chain in a tunnel enclosed by extended loops, which enables processive action where multiple cellobiose units are cleaved off from the non-reducing end of the chain before enzyme detachment. The glycosidic bond is hydrolyzed by a single-step inverting reaction mechanism, i.e. single displacement nucleophilic substitution at the anomeric carbon by an activated water molecule. However, the identity of the Cel6A catalytic base that abstracts a proton to activate the water molecule has remained a mystery. Path-sampling offers the tools for uncovering the reactive potential energy surface of the catalytic event, thus providing a molecular-level understanding of the roles of each residue in the cleaving of the strong glycosidic bond. Our combined quantum mechanics and molecular mechanics (QM/MM) simulations reveal the key role of a water wire in the shuttling of a proton away from the active site to the putative base, as well as providing a means for catalytic rescue upon mutation of the putative base. In addition to providing an atomistic understanding of enzyme action, this model provides a tool for the rational design of more efficient enzymes for use in producing renewable chemicals and fuels from non-food biomass.

429344 Modeling Chiral Recognition Between Amino Acids and Vaterite Surfaces

Abstract:

Accurate models of molecular recognition between biomolecules and mineral surfaces help to understand biomineralization processes, treat biomineralization related diseases such as the osteomalacias, and develop novel synthesis techniques for next-generation materials. Single amino acids have demonstrated chiral recognition on mineral surfaces, and their study provides insight into the behavior of proteins and peptides on those same surfaces. A better understanding of molecular recognition between amino acid building blocks and prebiotic minerals may also elucidate mechanisms for the coevolution of life and minerals on earth. Here we demonstrate how computational modeling with the RosettaSurface algorithm, combined with experimental data from *in vitro* crystal growth experiments, is used to study chiral recognition between amino acids and mineral surfaces. We computationally docked amino acid enantiomers to an array of vaterite surface models in order to propose and test different mechanisms to explain experimentally observed chiral crystal morphologies. The energy landscape for an amino acid near an ionic crystal is extremely rugged, necessitating Monte Carlo sampling techniques capable of overcoming energetic barriers. The RosettaSurface algorithm overcomes these barriers by generating $O(10^5)$ independent structural models using docking, side chain packing, and energy minimization. We have also developed a method for reflecting chiral amino acids across crystallographic mirror planes using homogeneous transforms. Our simulations suggest that orientationally dependent hydrogen bonding between amino acids and mineral surfaces is critical to achieving chiral selectivity. In addition to providing a potential explanation for the observed chiral morphologies, our docking studies also offer insight into the structure of the vaterite mineral, a topic of heavy debate for half a century due to multiple conflicting interpretations of vaterite's diffraction data.

434461 Toward Structure Prediction and Design of Protein Glycosylation

Abstract:

Glycosylation plays a critical role in protein folding, activity, stability, and processing. Glycosylation often underlies differences in protein properties when expressed in different organisms or under altered conditions such as in tumor environments or under metabolic stress. Three-dimensional models of the molecular structure of glycoproteins would be useful for identifying the molecular origins of these functional differences. However, few structural tools exist for identifying low-energy conformations of carbohydrates or glycoproteins. We have created a general tool for modeling the three-dimensional structures of oligosaccharides, polysaccharides, and glycoproteins as part of the Rosetta biomolecular modeling package. Our tool samples ring, backbone, and side chain conformations. We employ a library of sugar monomers and a methodology for chaining, branching, and modifying the monomers to create complex sugar polymers covering most biologically and synthetically relevant chemistries. Initial tests validate the accuracy of the models through prediction of ring conformation population frequencies (benchmarked to quantum calculations and x-ray crystal structure propensities) and docking of oligosaccharides (benchmarked to x-ray crystal complex structures). Some early applications include structure-determination of highly glycosylated protein structures and mechanisms of glycosylation effects on carboxylesterase activity. Finally, we present possibilities for computational design of glycosylations.

437173 Multiscale Simulations Can Reveal the Effect of Ionic Liquids on Structure and Dynamics of Biomolecules

Abstract:

Rational design of unique solvents and surfaces holds great potential for providing new ways to use biomolecules in engineering applications. Computational models such as molecular dynamics (MD) hold great potential for connecting the atomic scale to the mesoscale for a wide range of problems such as biocatalysis or protein stabilization by ionic liquids (ILs). Unfortunately, severe computational restrictions often limit wide-ranging use of these tools to their full potential. New multiscale modeling algorithms that are based on MD have been developed that can overcome these challenges, dramatically increasing the computer's viability as a tool for computation-driven discovery. This talk will highlight how we are using simulations to understand the dominant driving forces that lead to unique orientation and conformation at the protein/IL interface. We will cover both examples of using classical all-atom MD simulations as well as the application of metadynamics to study the thermodynamics of various processes like protein folding, highlighting key findings applied to biomolecule structure, dynamics, domain rearrangement and protein folding. Systems discussed in this talk include both large (i.e, multidomain) proteins and small model peptides.

437174 Computer Simulations in the Massively Parallel Era and Beyond

Abstract:

Computer simulations are now crucial in every field of science. Researchers need flexible simulation, analysis, and visualization tools to complete their work unencumbered. Researchers also need fast tools to reduce the design, experiment, iterate latency. Single threaded performance hit a brick wall many years ago, so researchers can no longer just run old code on new hardware to run simulations faster. Moore's law still gives us double the transistors with each new generation (for now), and all those go into wider parallelism. So today, and for the foreseeable future, all codes must go parallel or be left behind. In this talk, I will give a perspective on massively parallel computing in the soft matter field - where it has been and where it is heading. And I will highlight my own research results where high performance computing was critical. I lead the development of HOOMD-blue, a general purpose particle simulation toolkit that supports molecular dynamics of point and anisotropic particles, and hard particle Monte Carlo. It uses massively parallel GPUs to execute fast, and scales to all 18,000 GPUs on the Titan supercomputer. I value usability just as much as performance, and will discuss how tools with a Python language interface allow researchers to quickly and easily develop complex workflows and combine simulation, visualization and analysis into a single cohesive whole.

437016 Entropic Control over Nanoscale Colloidal Crystals

Abstract:

Globally ordered colloidal lattices have broad utility in a wide range of novel optical and catalytic devices, for example, as photonic bandgap materials. However, the self-assembly of stereospecific structures is often confounded by defects. Small free energy differences different crystal polymorphs, making it difficult to produce a single morphology at will. Current techniques to handle this problem usually rely on energy minimization; many colloids have been computationally engineered with anisotropic pairwise interactions to achieve morphological control. However, the complexity of these designs often makes experimental realization difficult. In this presentation, I plan to summarize recent computer simulation work enabled by efficient GPU-centric codes on the effects of polymeric co-solutes on crystallizing colloidal suspensions. These can be used to direct colloidal structures by relying upon the polymer's entropic interactions resulting from the interplay between the polymer's internal degrees of freedom and the void structure of a material. This represents a novel design paradigm that has the potential to significantly simplify control over colloidal polymorphism. I will elaborate on how to rationally design the cosolute structure to thermodynamically stabilize a single desired polymorph in a binary mixture, and the consequences that thermal perturbations have on this effect. I will then offer insights into how to design temperature dependent cosolute "switches" that allow the stability of a polymorph to be controlled via†experimentally accessible parameters.

436845 Invited Talk: Materials Cartography- Mapping Chemical Design Pathways

Abstract:

This presentation discusses how by discovering and analyzing patterns or "shapes" (the topology) of data can allow us to extract new insights for materials discovery and design that hitherto had been impossible or very difficult through existing modeling and experimental methods in materials science. Topological data analysis and materials informatics can help uncover geometric structures behind large, diverse, noisy, unstructured datasets It will help us discover new knowledge without necessarily requiring an a priori hypothesis; identify groups or classifications of data that would otherwise be hidden; and uncover networks of information flow that elucidates the pathways through which heterogeneous and diverse data sets are connected. Using examples in materials chemistry, we propose a framework to formalize the concept of "Materials Cartography" to harness the tools of data topology and informatics to map the "shape" and thereby extract new insights in materials discovery and design.

432270 Materials Informatics Platform for Property Prediction

Abstract:

The Materials Genome Initiative is transforming Materials Science into a data-rich discipline. These developments open exciting opportunities for knowledge discovery in materials databases using informatics approaches to inform the rational design of novel materials with the desired physical and chemical properties. Statistical and data mining approaches have been successfully employed in both chemistry and biology leading to the development of cheminformatics and bioinformatics, respectively. However, until recently their application in materials science has been limited due to the lack of sufficient body of data.

In this work we showcase a pilot materials informatics platform capable of (i) instantaneously query and retrieve the necessary material information in the desired form, (ii) identify, visualize and study important data patterns, and (iii) generate experimentally-testable hypotheses by building predictive Quantitative structure-activity relationship (QSPR) models based on materials' characteristics. Our computational approach relies on cheminformatics methodologies that one of our groups has developed and employed successfully to enable rational design of organic compounds with desired properties (e.g., drug candidates).

Specifically, we posit that materials with similar structural, topological, and electronic characteristics are expected to have similar physical chemical properties irrespective of their formal composition. To enable uniform comparison of materials by their intrinsic properties, w*e will represent all materials uniquely by multiple numerical descriptors, or fingerprints. This representation will enable the use of classical cheminformatics and machine-learning approaches to mine, visualize, and model any set of materials as we demonstrated in our recent pioneering studies on materials cartography [1].*

Reference:

1. Isayev, O., Fourches, D., Muratov, E,N,. Oses, C., Rasch K.M., Tropsha, A., and Curtarolo, S. *Chem, Mater,, **2015**, In press. DOI:* 10.1021/cm503507h

411269 A Learning Framework to Accelerate First Principles Simulations

Abstract:

First principles quantum mechanics based modeling schemes, such as density functional theory (DFT), are powerful tools to study the static and dynamical evolution of processes (e.g., chemical reactions, phase transformations, transport). Nevertheless, such methods have several practical drawbacks. For instance, owing to the computational expense, a typical DFT-based simulation can only handle system sizes of at most 100s of atoms or span timescales of the order of just picoseconds (with a maximum reachable timescale of about a nanosecond). Here, we show that DFT-based simulations can be significantly accelerated, by using machine-learning methods to make rapid high-fidelity predictions, based on past data or knowledge.

For many atomistic simulations, e.g. molecular dynamics, geometry optimization, identifying reaction barriers, materials properties, etc., determining the force on an atom is key. If we are able to rapidly estimate (with acceptable accuracy) the atomic forces of a new configuration, given past similar configurations (determined, say, using DFT), then we can significantly speed up the simulations. The 'training' underlying this force prediction capability requires a critical amount of prior information or results, adequate descriptors (or fingerprints) that uniquely represent our configurations, and a suitable measure of (dis)similarity between configurations. We show here that such training, prediction, and consequently, the acceleration of DFT-based simulations, are indeed possible. Illustrations of this new development are made for the case of the self-diffusion of vacancies in bulk Al, the diffusion of an Al adatom on an Al surface, finding optimal geometries of bulk Al structures with more than 100s of atoms, identifying reaction barriers for simple elementary processes, and lastly, determining thermodynamic properties of bulk Al. Such a framework lays the foundations, for understanding the behavior of chemical reactions or materials at increased time and length scales, all at first principles accuracy, a capability that is well beyond the current realm of first principles methods.

431228 Strategies for Efficiently Exploring Structure-Property Spaces Via High-Throughput Screening of Hypothetical Materials

Abstract:

Exploring structure-property spaces is an integral, but challenging, step in application-driven material design. The practically infinite number of possible materials makes it difficult to map a meaningful fraction of the structure-property space. Experimental screening of *real* materials consumes significant time and physical resources, and although computational screening of *hypothetical* materials allows one to map a larger space in less time, the number of possibilities is still too large for a brute force approach in either case. In this work, we describe an efficient method for mapping the physical limits of a structure-property space by randomly-generating an initial population of hypothetical materials, simulating their properties, and then generating subsequent populations by mutating those materials whose properties are rare. Our initial application area to test the method focused on methane adsorption in porous media. We considered our *finite* exploration of the structure-property space sufficiently representative of the *infinite* space, when adsorption data varied smoothly as a function of structure (e.g., void fraction, pore size, volumetric surface area) everywhere in the structure-property domain (except at the boundaries).

Initial populations are generated by populating unit cells with randomly positioned "pseudo" atoms, whose Lennard-Jones force field parameters are also randomly chosen. Methane adsorption at 35 bar and 298 K is simulated for each material using classical grand canonical Monte Carlo simulations. In addition, structure properties such as void fraction and surface area are also calculated for each material. Following an analysis of the initial population, we identified structure-property combinations that were rare, and then mutated materials with those properties to generate new populations. The mutation script perturbs all of a material's defining parameters (atom site positions, Lennard-Jones parameters, etc.) to a degree dictated by a strength parameter. Once the structure-property space has been "filled," which we define as the point where every structure-property combination is equally represented, we stop generating new materials (i.e., the method has converged). In this work, we investigate the impact of adjusting the mutation strength parameter on the speed of convergence. The method developed in this work provides an efficient way to map an entire structure-property space using only a small, finite, number of hypothetical materials. This work will help guide the design of materials for methane storage and potentially other gas-related applications in the future.

430712 Application of a Genetic Algorithm to Screen Metal-Organic Frameworks: Pre-Combustion CO2/H2 Separation

Abstract:

Metal-organic frameworks (MOFs) are novel porous materials consisting of organic and inorganic building blocks, whose combinatorial possibilities give rise to a practically unlimited number of possible structures. Due to the large number of possible MOFs, high-throughput computational screening has been used in the literature to identify high-performing MOFs for particular applications of interest, such as methane storage and delivery. Computational screening has also provided useful structure-property relationships that can guide experimental discovery efforts. However, current approaches are "brute-force", where full molecular simulations are carried out to calculate the property of interest for a very large number of structures that have been generated or mined from the literature. This approach may not be suitable, however, for complex systems or for very time-consuming simulations because of the prohibitively large computational cost associated with such simulations.

In this work, we have developed a Genetic Algorithm (GA) to identify high-performing MOFs for pre-combustion CO_2/H_2 separation from a database of ~137,000 hypothetical MOFs [1]. Firstly, grand canonical Monte Carlo (GCMC) simulations were carried out to calculate CO_2/H_2 uptakes under selected operation conditions for a "genetically diverse" initial generation of 100 hypothetical MOFs. Secondly, genetic operations were carried out on the preceding generation of structures to create a new one, for which GCMC simulations were carried out as well. This procedure was continued until we reached 10 generations.

Our results demonstrate that the developed GA is a robust method that can find high-performing MOFs based on different performance metrics such as CO_2/H_2 selectivity and CO_2 working capacity in a much more computationally efficient fashion than the brute force method. One of the top MOFs identified by the GA was synthesized and activated, with the experimental CO_2 and H_2 adsorption values in good agreement with the molecular simulation results. Moreover, the predicted MOF has the highest CO_2 working capacity (~ 6 mol/kg) among reported MOFs and zeolites in the literature under the same operating conditions [2].

#: Authors contributing equally

References

[1] C. E. Wilmer, M. Leaf, C.Y. Lee, O.K. Farha, B.G. Hauser, J.T. Hupp, and R.Q. Snurr, "Large-scale screening of hypothetical metal-organic frameworks", Nature Chemistry, **2012**, 4, (2), 83 – 89.
[2] Z. R. Herm, J. A. Swisher, B. Smit, R. Krishna, and J. R. Long, "Metal-Organic Frameworks as Adsorbents for Hydrogen Purification and Precombustion Carbon Dioxide Capture", J. Am. Chem. Soc., **2011**, 133, (15), 5664 – 5667.

435035 Computational Screening of MOFs with Open Metal Sites

Abstract:

Computational Screening of MOFs with Open Metal Sites

Emmanuel Haldoupis[1], Konstantinos D. Vogiatzis[1], Laura Gagliardi[1], J. Ilja Siepmann[1,2]

[1]Department of Chemistry and Chemical Theory Center, University of Minnesota, Minneapolis, Minnesota 55455, United States

[2]Department of Chemical Engineering and Materials Science, University of Minnesota, 421 Washington Avenue SE, Minneapolis, Minnesota 55455, USA

Metal-organic frameworks (MOFs), among other nanoporous materials, are of potential utility for many energy-relevant or technologically important processes, such as adsorption and catalysis. Specifically undercoordinated metal centers, often found in MOFs, can interact strongly and selectively with certain guest molecules. These strong interactions may involve partial bond formation leading to highly selective separations compared to MOFs with fully coordinated metal centers and allowing for catalytic activation for various reactions. Computational high-throughput screening of materials can be used to assess the performance of thousands of materials. However, correctly modeling the strong interaction of guest molecules with open metal centers requires the use of expensive quantum chemical methods and reduces the number of materials that can be examined. To mitigate this challenge we have developed a computational tool that analyzes the atomic coordinates of MOFs, identifies the presence of patterns corresponding to open metal sites with sufficient guest accessibility and, hence, reduces the number of materials that need to be examined using more expensive computational methods. Recently a database containing more than 5000 computation-ready experimental (CoRE) MOF structures with a pore limiting diameter of at least 2.4 Å was reported [1]. After screening this database and identified all MOFs with open metal sites we concentrate on MOFs containing Mg and Fe open metal sites. We identify 51 MOFs with unsaturated Mg sites and assessed their CO_2/N_2 adsorption selectivites using periodic DFT calculations. In addition, 43 MOFs with coordinatively unsaturated iron centers are identified and examined by means of electronic structure theory for the stability of the highly reactive Fe(IV)-oxo intermediate.

[1] Y. G. Chung, J. Camp, M. Haranczyk, B. J. Sikora, W. Bury, V. Krungleviciute, T. Yildirim, O. K. Farha, D. S. Sholl and R. Q. Snurr, ``Computation-Ready, Experimental Metal-Organic Frameworks: A Tool To Enable High-Throughput Screening of Nanoporous Crystals", *Chemistry of Materials*, **2014**, 26, 6185-6192.

426209 Graph Based Technique to Quantify Charge Transport Characteristics of Heterogenous Morphologies

Abstract:

The nanomorphology of polymer blend thin films critically affects performance especially in electronic devices. However, many aspects of the underlying physics linking morphology to performance are still poorly understood. Furthermore, there is increasing evidence that atomic organization can hold the key to efficient charge transport within organic electronic devices. In this work, we take advantage of recent advances in molecular dynamic simulations and quantify atomic-scale morphological aspects of the thin films. Specifically, we present a graph-based technique that allows quantifying the point-cloud data in heterogenous morphologies. In our approach, we first convert the point cloud data from atomistic simulation into a labelled, weighted, undirected graph and then use standard graph-based algorithms to calculate and quantify morphology features. The conversion of the CGMD-data into a graph preserves all the topological and geometric information about the internal structure, and local connectivity between individual atoms/beads (along and across the polymer chains). Our method provides hierarchical information about the charge paths that a hole/electron needs to take to reach the electrode (path length, fraction of intra-molecular hops, path balance). We showcase capabilities of our approach by analyzing coarse grained molecular simulations, and quantifying effect of various thermal treatment as well as electrode materials on the P3HT:PCBM blend.

434314 Deconstructing Detailed Reaction Mechanisms to Identify Species and Mine for Data

Abstract:

Detailed kinetic models have become integral to combustion research over the last 40 years. The latest models can explain many complicated phenomena and offer accurate simulations for novel engines and fuels. These models can be very large (eg. the LLNL model for 2-methylalkanes has over 7,000 species and 30,000 reactions) and there are now dozens of these large published models. Unfortunately, these ever-proliferating detailed kinetic models are usually incompatible and inconsistent, are seldom compared directly, and often contain undetected mistakes.

The usual publication format remains a "Chemkin file" for use in compatible simulation tools. This format, devised in the 1970's when input was limited by the width of 80-column punch-cards, forces model-builders to abbreviate species' names, thereby losing their chemical identity, and to discard other metadata. The main challenge in comparing these models is in recognizing, for example, that the name "C3KET12" in one model represents 1-hydroperoxypropan-2-one, which another research group may have named "CH3COCH2O2H" in a different model.

The clues available to determine the molecule corresponding to a given nickname are: the name itself (often cryptic and sometimes misleading), thermochemical data (usually estimated), and a list of reactions (usually incomplete) in which the species participates, connecting it to other species in the model (that are also initially unknown).

We have developed an algorithm and tools to facilitate the identification of chemical species in a kinetic model "Chemkin file", and then to allow comparison of the models. The tools are built on top of the open source Python version of Reaction Mechanism Generator software (RMG-Py)[1], originally designed to create detailed kinetic models of its own. By comparing reported reactions with its own predicted reactions between already-identified species, it is able to propose new species and eliminate unlikely matches. A web-based user interface allows a team of humans to quickly review the evidence and confirm or block the proposed matches.

We will present how the tool works, opportunities for improvement, and some findings from analyzing recent publications in the combustion kinetics literature. This material is based upon work supported by the National Science Foundation under Grant No. 1403171.

[1] Green, W.H., Allen, J.W. et al. RMG – Reaction Mechanism Generator, Python Version. http://rmg.mit.edu, 2013.

419965 Molecular Theory for the Hydrophobic Interactions and Absorption of Span80 at a Squalane-Water Interface

Abstract:

Molecular theory for the Hydrophobic Interactions and Absorption of Span80 at a Squalane-Water Interface

M. I. Chaudhari, S. B. Rempe, D. Asthagiri, L. Tan, L. R. Pratt

Abstract: The principal result of LMF theory is outlined, then tested by obtaining radial distribution functions (rdfs) for Ar atoms in water, with and without attractive interactions distinguished by the Weeks-Chandler-Andersen (WCA) separation. Change from purely repulsive atomic solute interactions to include realistic attractive interactions diminishes the strength of Ar-Ar hydrophobic bonds. Since attractions make a big contribution to hydrophobic interactions, Pratt-Chandler theory, which did not include attractions, should not be simply comparable to computer simulation results with general physical interactions, including attractions. The Ar-Ar osmotic second virial coefficients become more attractive with increasing temperature below T = 360K. Ultimately, LMF theory does not accurately describe the numerical results for the effects of solute attractive forces on hydrophobic interactions in this case. We extend this work to investigate the adsorption/desorption behavior of surfactant at the oil/water interface. According to recent experimental work, TWEEN80 shows an irreversible adsorption behavior at a squalane/water interface when the surface tension is above 32mN/m and an alternative partial reversibility when the surface tension is below this critical value. We use molecular dynamics to study of the squalane/water/surfactant interface, substituting SPAN80 for TWEEN80 as an initial step for simplicity. We apply parallel tempering calculations to obtain the surface tension of this interface. The temperatures for the parallel tempering simulations range from 260K to 450K in order to achieve an exchange probability between neighboring temperatures around 20%. We then further exploit a windows sampling strategy to study the free energy variation for desorption of a SPAN80 molecule from the interface at 300K. Preliminary pulling results suggest that the observed irreversibility of the adsorption/desorption is due to kinetic sluggishness of the SPAN80 tail motion in the squalane phase.

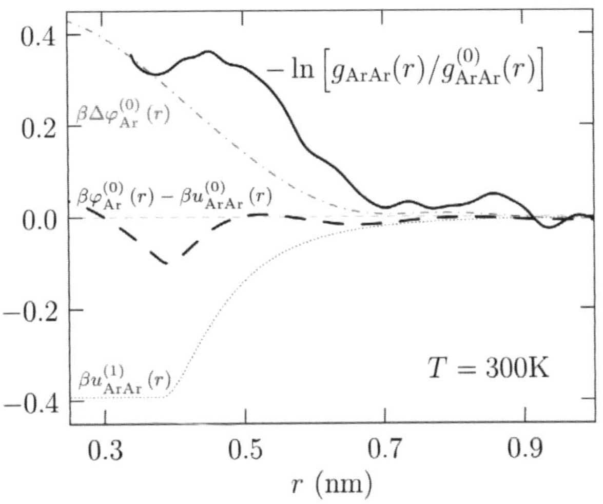

Fig. 1: Test of the LMF theory. The net result $\beta\varphi^{(0)}_{\mathrm{Ar}}(r) - \beta u^{(0)}_{\mathrm{ArAr}}(r)$ (black-dashed) is the sum of the red and blue curves; $\beta\Delta\varphi^{(0)}_{\mathrm{Ar}}(r)$ (red-dotdashed) is the solvent contribution. $-\ln\left[g^{(0)}_{\mathrm{ArAr}}(r)/g_{\mathrm{ArAr}}(r)\right]$ differs from $\beta\varphi^{(0)}_{\mathrm{Ar}}(r) - \beta u^{(0)}_{\mathrm{ArAr}}(r)$ both in contact and solvent separated configurations.

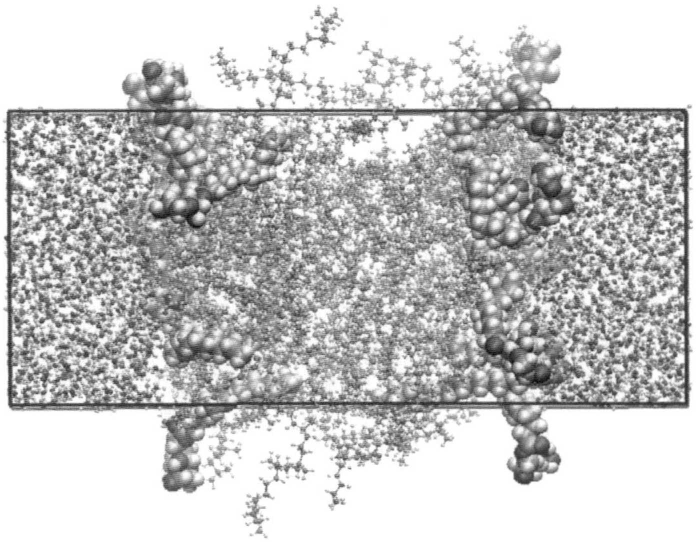

Fig. 2: Snapshot of a water/squalane/SPAN80 configuration. Water fills two sides of the picture with the squalane in the middle. Ten SPAN80 molecules occupy the squalane/water interface at each side.

431135 Molecular Modeling of Select Organic Molecules at the Air-Water Interfaces

Abstract:

A large sector of industrial chemicals are amphiphilic organic compounds. When dissolved in water, amphiphlies are attracted to the air-water, i.e. nonpolar-polar, interface. The hydrophilic or polar head group of the compound is attracted by hydrogen bonding to water while the hydrophobic or alkyl chain group is attracted to the nonpolar air. In the concrete industry, amphiphilic compounds are used to reduce shrinkage cracking in portland cement concrete by reducing the surface tension of the pore water within the cement microstructure. The cement pore water contains sodium, potassium, calcium, sulfate and to a lesser extent other ions. Assembly and concentration of the amphiphilic compound at the air-water interface in the presence of various ions is critical to the understanding of compound effectiveness. To elucidate the behavior of select compounds, a combined experimental and computational approach was used. This study summarizes ongoing molecular dynamics simulations and experiments using three compounds: butyl glycolate, 2-(2-butyoxyethoxy) ethyl acetate, and hexylene glycol. Molecular dynamics simulations were performed to simulate the interactions of the test compounds pure water and simulated pore solution. Interactions between various groups were modeled using the COMPASS force field. In addition, surface tension experiments were also used to elucidate the process and reconcile model predictions and experimental observations.

431575 Molecular Dynamics Simulation of Amphiphilic Graphene Oxide As a Tunable Colloidal Surfactant in Oil/Water Mixtures

Abstract:

Recently, the amphiphilic nature of graphene oxide (GO) has been shown to provide a significant advantage for its use as an efficient colloidal stabilizer and emulsifier in oil/water mixtures. The GO emulsification can be tuned by factors such as degree of oxidation, size of the graphene sheets, and pH level of the medium. In spite of several reports pertaining to GO as a tunable surfactant, a deep understanding of the emulsification process is still missing. Hereon, we study the effects of above-mentioned factors on the GO amphiphilicity using molecular dynamics (MD) simulation in BIOVIA Materials Studio software with COMPASS force field. A series of toluene/water/GO systems with varying GO sheet sizes, O/C ratios (surface oxygen densities), and pH levels are geometry-optimized. Then we run dynamics simulations using the NVT ensemble for 8 ns and analyze the equilibrated systems. The results of this study provide a more comprehensive picture of how to tune the GO emulsification for oil/water mixtures and the field of stimuli-responsive stabilizers, which are of great importance in crude oil extraction, recycling petroleum industry effluents, and pharmaceutical industries.

435347 Dynamics of Photo Switchable Patchy-Charged Surfaces

Abstract:

Tuning surface chemistry to achieve selective material properties has been an object of research for decades. Subtle variations in chemical environment at the surface of particles and materials often result in drastic changes in surface reactions, self- assemblies, surface toxicity, and more. Fine-tuning surface chemistry of materials (and/or nano particles' surfaces) in response to environmental signals could be an efficient way to dynamic surfaces capable of adjusting their surface properties and also manipulating material objects resting on such surfaces. Change of the surface properties could be triggered by alteration or gradient in chemical potential, external fields such as electrical or magnetic fields, light, or other environmental stimuli. Potentially applications of such smart surfaces might include: transferring matter (at different scales), capture and release of ions, self-cleaning materials, "smart" delivery carriers, etc. As an example, nanoparticles functionalized by photoactive ligands might undergo conformational changes upon light irradiation ultimately altering surface properties. Careful selection of grafted functional groups at the correct ligand density could yield photoactive surfaces of the particles having dynamic surface charge density or hydrophobicity. Since the relaxation time of light-induced processes is fairly short, such photoactive particles could be applied to trap and release ions or other charged moieties. In this research, Azobenzene invariant molecules are applied as to produce photo-induced switchable surface properties upon emitting *uv* light. Molecular dynamics simulations and quantum mechanical calculations were applied to design the optimal structures leading to systems with stable/metastable interchangeable properties.

432215 Contact Charging Between Insulators: A Theoretical Study

Abstract:

Triboelectric charging – the electrostatic charge that develops on surfaces when they are rubbed together – is ubiquitous, and has important consequences in many industries and many natural phenomena. However, the mechanism of the charging process remains very poorly understood. For example, it is not yet known what type of species are transferred during the charging process (electrons or ions), and there is no way to predict which surface charges positive and which charges negative. To give insight, we use a combined theory and experiment approach. We address a simple system – well-defined crystallographic faces of single-crystal quartz (SiO_2) and sapphire (Al_2O_3) – so that high level theory can be applied. The simulations use first-principles molecular dynamics, where density functional theory is used to determine the electronic structure of the system and the interatomic forces. Both dry and wet surfaces were investigated, where wet surfaces take into account the effects of ambient humidity. For dry situation, our results show that an electron transfer mechanism would cause sapphire to charge positively and quartz to charge negatively, in agreement with results from the experiments. This direction of charge transfer is shown to correlate with electrostatic potential energy on both surfaces. We also carry out simulations to with water, to address an ion transfer mechanism. Dissociation and interaction of water has been observed, and the trajectories of the hydroxyl ions (OH^-) and hydrogen ions (H^+) are examined. Additional simulations have addressed the interaction of quartz and sapphire with periclase (MgO).

415323 Solvent Effects on the Properties of Electric Double Layers with Surface Charge Regulation

Abstract:

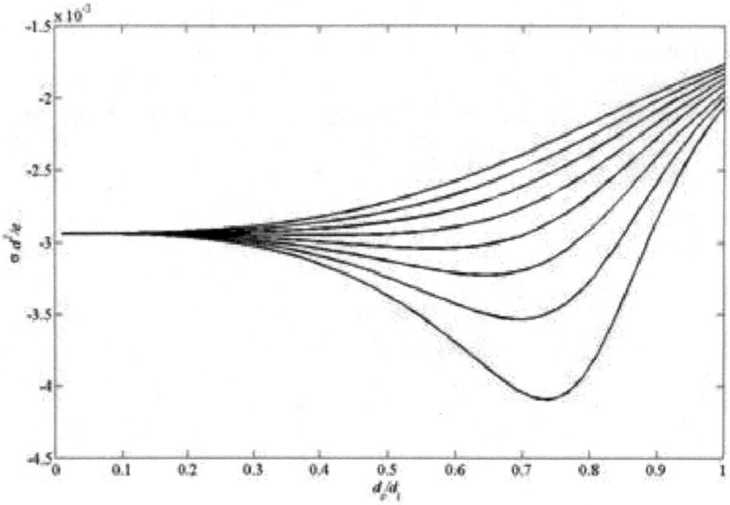

The thermodynamically consistent representation of the charge at the surface of an electric double layer is given by the chemical equilibrium between the groups attached at the interface and the potential determining ions in the solution (i.e., charge regulation). We report that this surface chemical equilibrium is strongly coupled to the precise molecular structure of the solution in near the charged interface that explicitly includes the solvent contribution, also known as a "civilized model". Our analysis uses classical density functional theory to obtain the molecular and ionic structure of an electric double layer with surface charge regulation. The model suggests that the excluded volume of the solvent is the main factor that couples the solution structure to the surface chemistry, although attractive interactions may also play a role. The Figure shows how the surface charge density depends on the relative size of the solvent molecules in a Lennard-Jones type of electrolyte solution. The different curves (top to bottom) are for increasing values of the Lennard-Jones energy. The conclusion from our work is that an explicit account of the solvent effects is necessary for quantitative description of an electric double layer with surface charge regulation.

429795 Multiscale Modeling of the Electrode/Electrolyte Interface Using Charge Optimized Many Body (COMB3) Potentials

Abstract:

MULTISCALE MODELING OF THE ELECTRODE/ELECTROLYTE INTERFACE USING MODIFIED CHARGE OPTIMIZED MANY BODY (ECOMB3) POTENTIALS

Sneha A. Akhade[1]*, Andrew Antony[2], Tao Liang[2], Michael J. Janik[1], Janna M. Maranas[1] and Susan B. Sinnott[2]

[1]Department of Chemical Engineering, Fenske Laboratory, Pennsylvania State University, State College, PA 16802

[2]Department of Materials Science & Engineering, 549 Gale Lemerand Drive, Gainesville, FL 32611, University of Florida

*Email: saa243@psu.edu

Present day advances in computer simulation have greatly accelerated material discovery and the investigation of bulk and surface properties that are relevant to the reactivity and performance of the material. Electrocatalytic systems present unique challenges to atomistic and electronic structure modeling due to the presence of an electrode/electrolyte interface and the integration of electrical currents and chemical reactions.

Our objective is to develop robust methods that can provide a reasonable atomistic description of the electrode/electrolyte interface that can be used to investigate the complex underlying interfacial chemistry. We combine Density Functional Theory (DFT) and a force field (FF) based classical approach to construct an atomistic modeling tool to consider this interface. Two key features of the electrochemical interface are incorporated - metal polarizability and the interfacial electrolyte chemistry in the presence of a potential-controlled field. In our previous study [1], we employ a reactive modified central force-field (mCF) model that uses simple point charges with pairwise interactions to model reactive water that can dissociate under the influence of potential controlled metal electrode. The electrodes are held at constant potential by using a charge fluctuation approach called Electrode Charge Dynamics (ECD) [2]. However the mCF model assigns a fixed charge/oxidation state to the water species and the ECD scheme does not permit charge exchange between the electrode and the electrolyte. In this work, we report third-generation Charge Optimized Many Body potentials (COMB3) [3] in conjunction with an electrode charge equilibration scheme (ECOMB3) for modeling the electrified interface. This reactive FF model advantageously allows for a multi-body charge equilibration within a constant potential controlled environment. Our results compare and test the efficacy of the two models using molecular-scale tools that are developed to probe the electrochemical and interfacial properties and provide a reliable atomistic description of the complex electrode/electrolyte interface.

1. Yeh, K.-Y., M.J. Janik, and J.K. Maranas, *Molecular dynamics simulations of an electrified water/Pt(111) interface using point charge dissociative water.* Electrochimica Acta, 2013. **101**(0): p. 308-325.

2. Guymon, C., et al., *Simulating an electrochemical interface using charge dynamics.* Condens. Matter Phys, 2005. **8**: p. 335-356.

3. Liang, T., et al., *Classical atomistic simulations of surfaces and heterogeneous interfaces with the charge-optimized many body (COMB) potentials.* Materials Science and Engineering: R: Reports, 2013. **74**(9): p. 255-279

426764 Surface Reaction, Solvent Inhomogeneity, and Ion Transport in Electric Double Layers: Predictions from a Classical Density Functional Theory

Abstract:

Electric double layers (EDLs) are formed at the interface of a charged fluid in contact with an electrified surface or an electrode leading to drastic changes in the fluid structure and local thermodynamic properties. Whereas conventional theories of EDLs are often concerned with the local inhomogeneity in terms of ionic distributions near a planar surface, the simplistic model is insufficient to capture many important properties relevant to emerging electrochemical processes for energy storage and conversion. In this talk, we present a few recent applications of the classical density functional theory (CDFT) that account for not only ionic excluded volume effects and electrostatic correlations ignored in conventional theories of EDLs but also electrode geometry and pore size distributions, the local solvent and dielectric inhomogeneity, surface reactions, and hydrodynamic effects. The CDFT calculations will be illustrated with practical examples for unusual ion transport in nanopores, optimization of capacitive energy extraction processes, design of nanostructured electrodes for flow supercapacitors and pseudo-capacitors. The theoretical predictions are validated with experiment results and provide insights for novel electrochemical behavior that are not captured in conventional EDL theories.

425274 Molecular Simulation of Adsorption and Separation of Pure Noble Gases and Noble Gas Mixtures on Single Wall Carbon Nanotubes

Abstract:

Grand Canonical Monte Carlo simulations were performed to systematically study the adsorption and separation of noble gases on single wall carbon nanotube (SWCNT) bundles. Pure noble gases, as well as binary and ternary mixtures, were simulated in carbon nanotube systems under various conditions. Adsorption data was collected at 100 K and 300 K over a wide range of pressures. Carbon nanotube bundles present distinct adsorption capacities towards different noble gases. In particular, larger and heavier noble gases are easier to be adsorbed at low pressure, while lighter atoms with smaller sizes can be better stored at high pressure. For noble gas mixtures with equal molar loadings, selective adsorption was observed and the selectivity inverted at different pressure ranges according to the choice of mixtures. Furthermore, the influence of loading proportion of the components on the adsorption behavior was investigated by varying the loading partial pressure in binary mixtures from 1% to 99%. Adsorption selectivity was only affected in sub-saturation conditions and no obvious difference in adsorption amount was observed once saturation was reached. The results suggest gas-CNT interactions dominate the adsorption selectivity at unsaturated conditions, whereas the entropic effect plays a more important role after saturation.

433421 Molecular Dynamic Simulation of Clay Particles Swelling

Abstract:

Clay minerals, such as smectites and mixed-layer illites, can expand in volume up to 20 times their original volume through adsorption of layers of water between their unit cells. This is a very important phenomenon as swelling clays reduce formation permeability and porosity by peeling off the pore surfaces and plugging pore throats, not by reducing porosity alone.

In an attempt to understand the clay swelling process due to shear, we are using MD simulation to study the swelling and shrinkage of clay particles. A slit pore, the space between two parallel surfaces, is used. In order to show the coupling between swelling and shear in liquid-saturated clay pores, we use Montmorillonite structure and SPC/E model of water molecules. First, the consistency of the simulations using thermodynamics and direct routes is validated when dealing separately with swelling and shear. Then, the coupling between swelling and shear is explored by displacing the solid walls in one direction while letting them move freely on the other.

Our results so far indicate that shear can induce swelling and vice versa, due to the confined fluid phase structure. It implies that the response to a variation in the external load can be a combination of volumetric and shear deformations, because of the presence of the fluid.

We are also exploring the behavior of the pore system induced by solid walls moving at a constant velocity. Interestingly, when the wall velocity exceeds the swelling velocity, the instantaneous states of the system are no longer at equilibrium and the averaged pore width slightly increases with increasing shear rate.

We are also studying the effect of NaCl concentration on the swelling process. So far, we have found that concentration of Na affects the swelling process, such that as the concentration increases, the spacing between two layers of clay decreases.

Disruption of Ice Adhesion Through Surface Functionalization

Ben Schweitzer, Kevin Hadley, Nick McDougall, Joseph Smith, Christopher Wohl, Richard Kreeger, and Jose Palacios

Background: Motivation

- Aviation

National Transport Safety Board (2011). Lessons from Icing Accidents and Incidents.

Background: Current Research

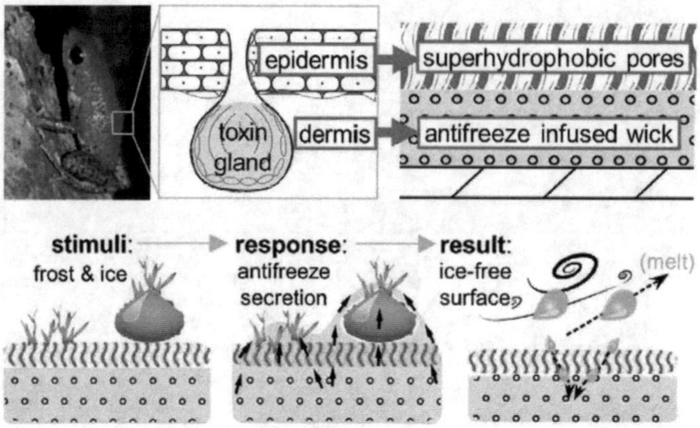

Farhadi, S., et al. (2011). "Anti-icing performance of superhydrophobic surfaces." Applied Surface Science **257**(14): 6264-6269.
Jung, S., et al. (2011). "Are Superhydrophobic Surfaces Best for Icephobicity?" Langmuir **27**(6): 3059-3066.
Sun, X., et al. (2015). "Bioinspired Stimuli-Responsive and Antifreeze-Secreting Anti-Icing Coatings." Advanced Materials Interfaces **2**(5).

Our Approach

Biomimetic Surface

- Clathrate Hydrates
- Antifreeze Proteins

Form favorable hydrogen bonds with water

- Steal hydrogen bonds from water
- Disrupt the ordered hydrogen bonding network
- Increases the free energy barrier to add more ice layers.

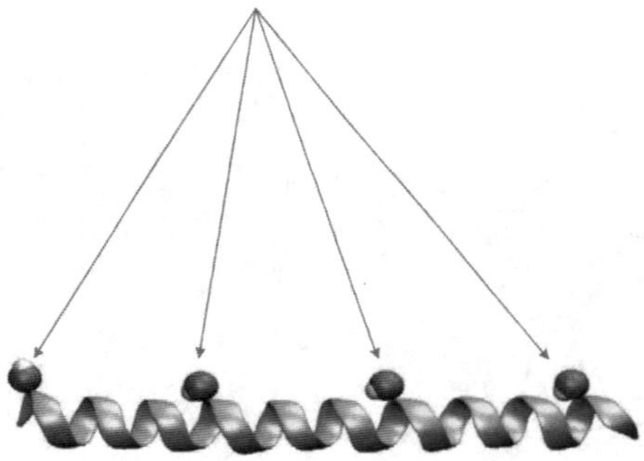

Our Approach

Biomimetic Surface

- Clathrate Hydrates
- Antifreeze Proteins

Form favorable hydrogen bonds with water

- Steal hydrogen bonds from water
- Disrupt the ordered hydrogen bonding network
- Increases the free energy barrier to add more ice layers.

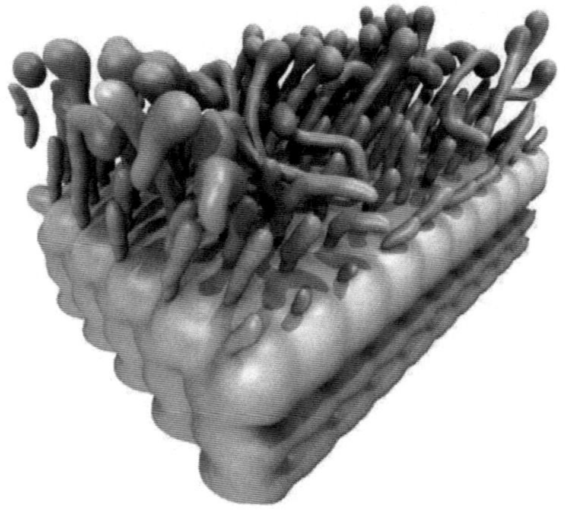

Methods

Modifiers

Varied carbon chain lengths

- 12 carbons for long chains
- 7 carbons for short chains
- Randomized chain lengths

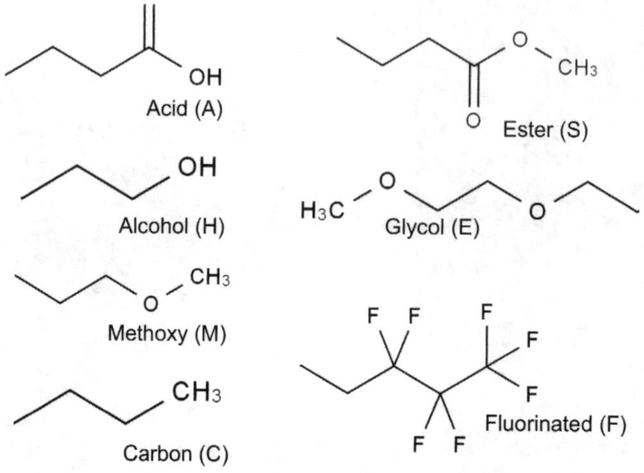

Structure of Ice

- Quasi Liquid Layer (QLL)
  - Forms at the interface of ice and the surface
  - Lowers the free energy of the ice-surface Interface
- Quantified by the F4 Structure factor
  - Dihedral angle between two water hydrogen bonds
  - $F_4 = \langle \cos(3\emptyset) \rangle$
  - -0.4 is Ice and -0.04 is liquid water
  - -0.29 is the interface between
  - Moving average Z axis

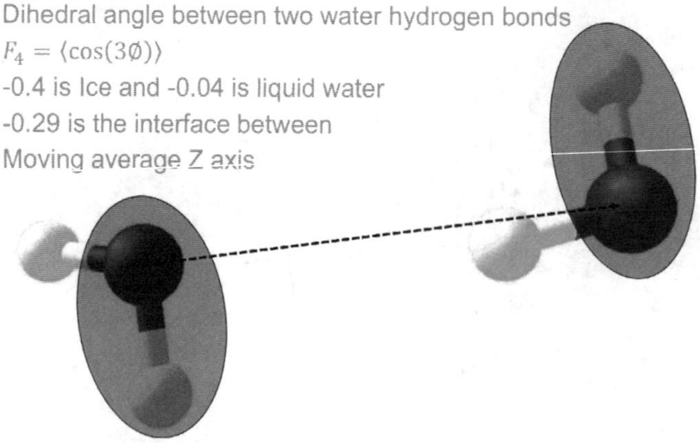

Methods

- Equilibrate the Surface
  - Heated to 300K
  - Cooled to 230K
- Let the Ice Crystal Rest on the Surface
  - 50K
- Let the QLL Develop Naturally
  - 230K
  - 5ns

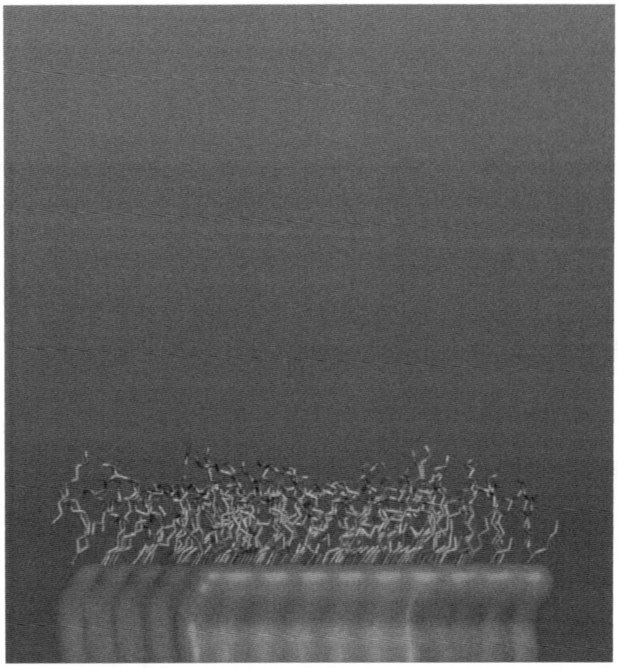

Methods

- Equilibrate the Surface
  - Heated to 300K
  - Cooled to 230K
- Let the Ice Crystal Rest on the Surface
  - 50K
- Let the QLL Develop Naturally
  - 230K
  - 5ns

Methods

- Equilibrate the Surface
  - Heated to 300K
  - Cooled to 230K
- Let the Ice Crystal Rest on the Surface
  - 50K
- Let the QLL Develop Naturally
  - 230K
  - 5ns

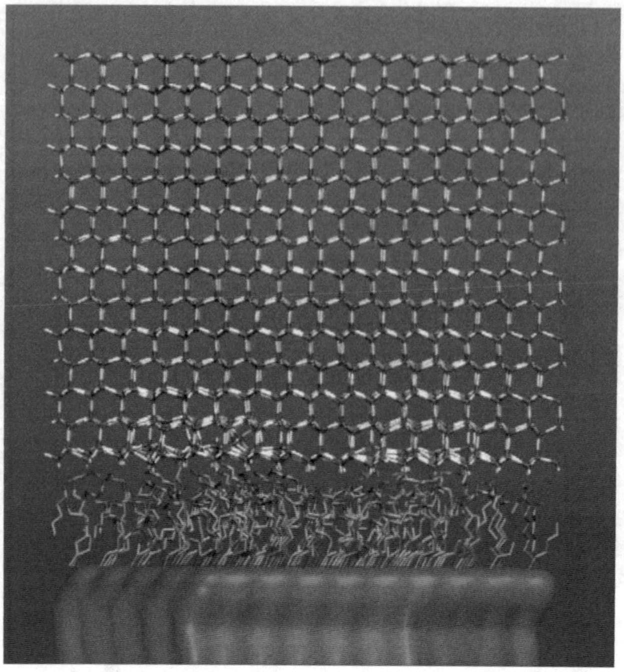

Methods

- Equilibrate the Surface
  - Heated to 300K
  - Cooled to 230K
- Let the Ice Crystal Rest on the Surface
  - 50K
- Let the QLL Develop Naturally
  - 230K
  - 5ns

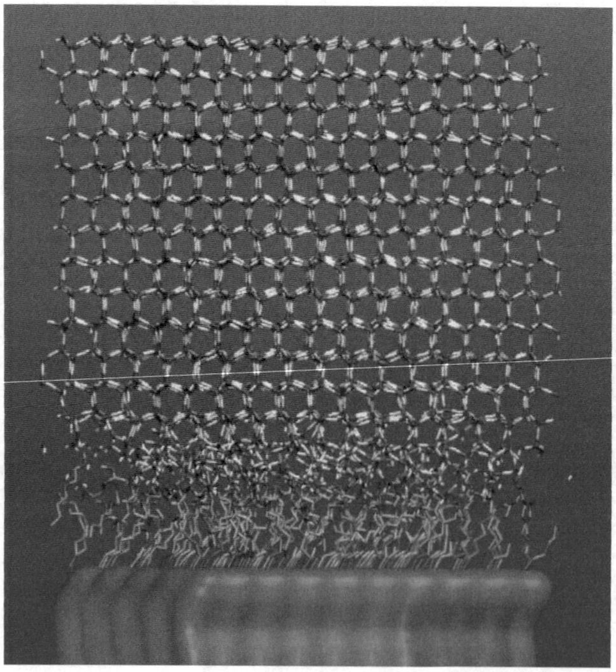

Simulation Details

- LAMMPS (1Feb2014)
- TIP4P2005 / ICE
  - Best Properties at 230K
  - Computationally Expensive
- Dreiding – Short range interactions
  - Atoms, Bonds, Angles, Dihedrals
  - Good Equilibrium Structures
  - United atom carbon chains
  - Save Computational time
- CHARMM – Long range interactions (Charges and LJ)

Simulation Results

Simulation Results

Moeities	
Carbon	C
Alcohol	H
Methoxy	M
Glycol	E
Acid	A
Fluorocarbon	F
Ester	S

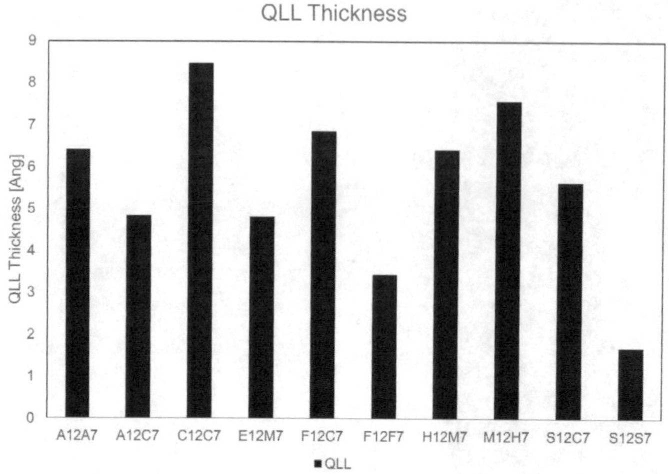

New Perspective

- Bulk
- Quasi Liquid
- Interface
  - Bulk Ice – QLL
  - Surface - QLL

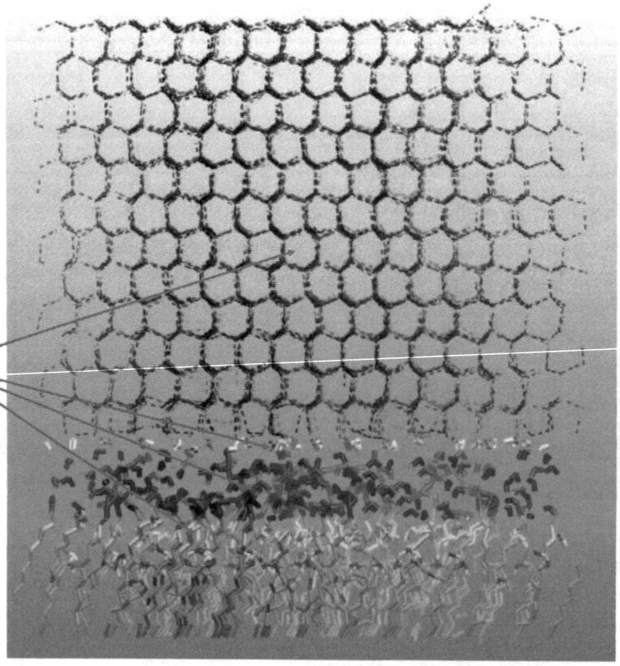

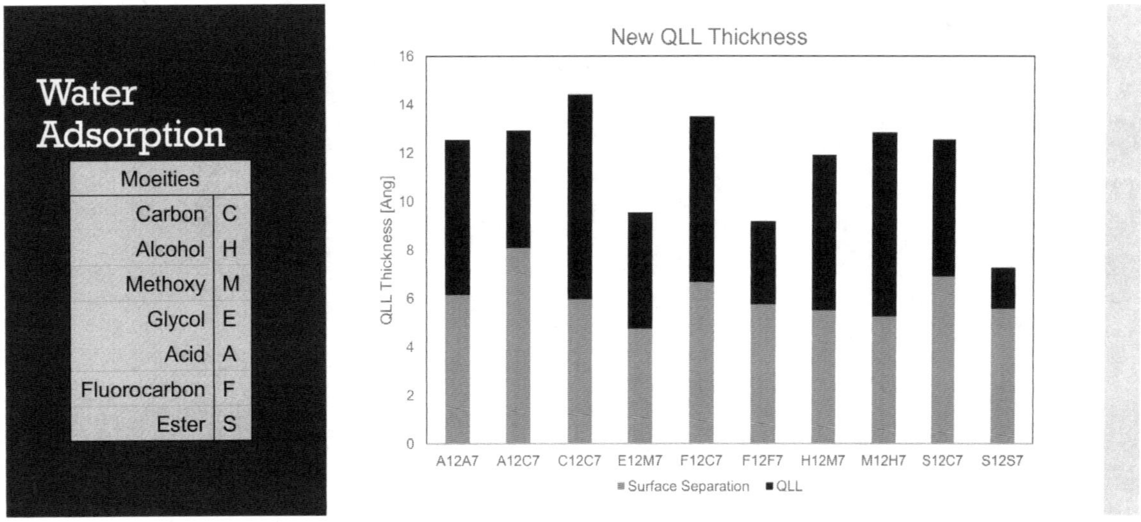
Water Adsorption
Moeities
Carbon C
Alcohol H
Methoxy M
Glycol E
Acid A
Fluorocarbon F
Ester S
New QLL Thickness
QLL Thickness [Ang]
0
2
4
6
8
10
12
14
16
A12A7
A12C7
C12C7
E12M7
F12C7
F12F7
H12M7
M12H7
S12C7
S12S7
Surface Separation
QLL

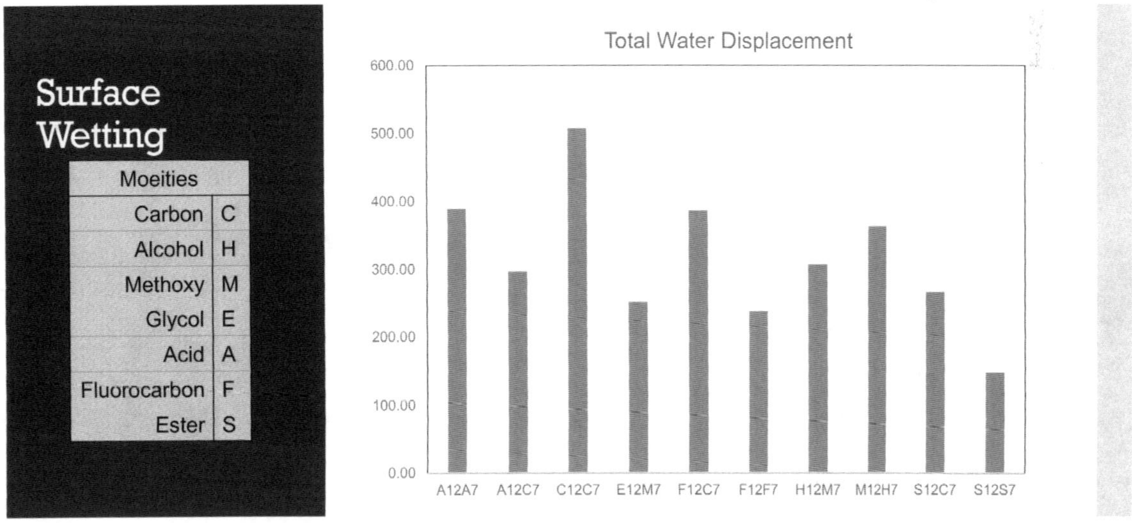
Surface Wetting
Moeities
Carbon C
Alcohol H
Methoxy M
Glycol E
Acid A
Fluorocarbon F
Ester S
Total Water Displacement
0.00
100.00
200.00
300.00
400.00
500.00
600.00
A12A7
A12C7
C12C7
E12M7
F12C7
F12F7
H12M7
M12H7
S12C7
S12S7

Surface Wetting

Moeities	
Carbon	C
Alcohol	H
Methoxy	M
Glycol	E
Acid	A
Fluorocarbon	F
Ester	S

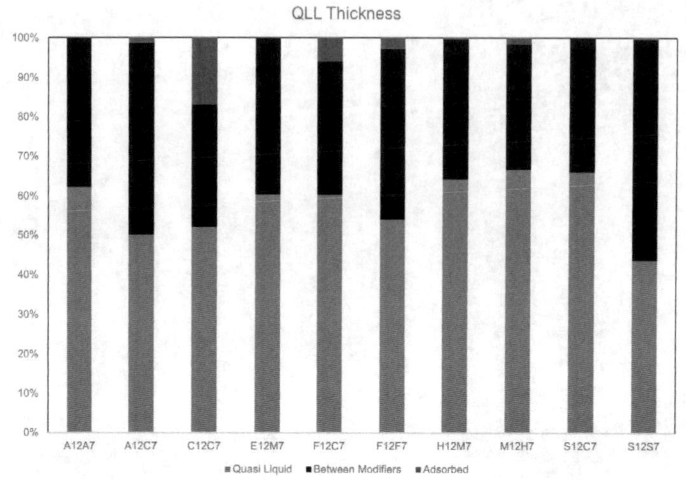

QLL Density

Moeities	
Carbon	C
Alcohol	H
Methoxy	M
Glycol	E
Acid	A
Fluorocarbon	F
Ester	S

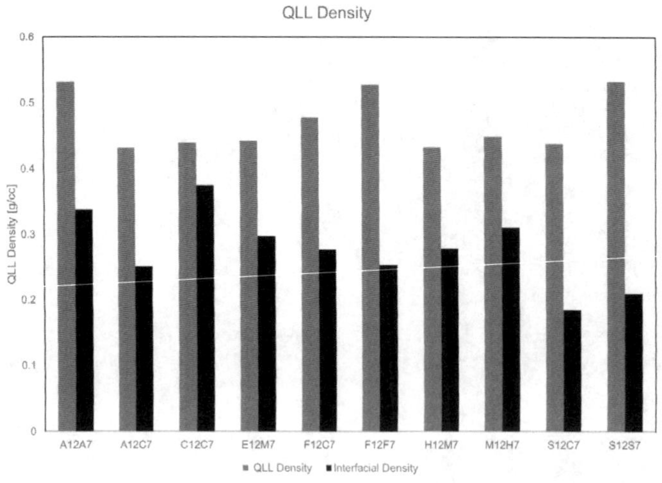

Templating Surfaces

- M12C7 and E12C7
  - Aliphatic Short Chains
- No QLL formation
- Template for ice formation
- Icephylic

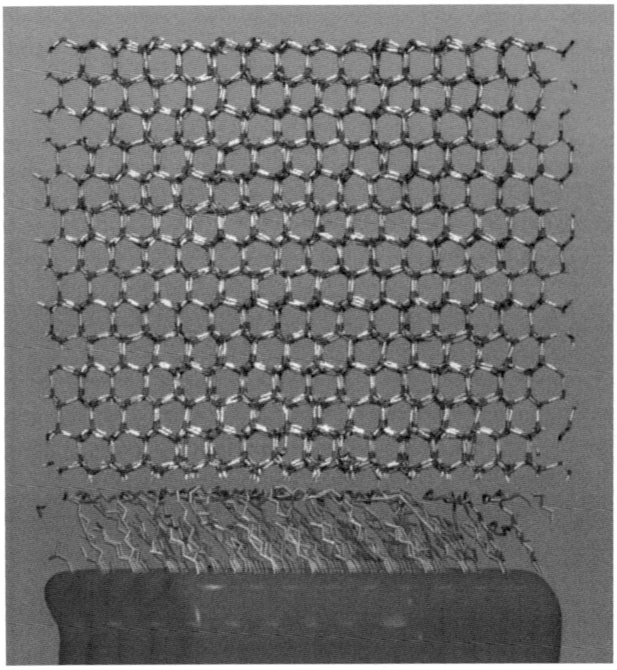

Templating Surfaces

- M12C7 and E12C7
  - Aliphatic Short Chains
- No QLL formation
- Template for ice formation
- Icephylic

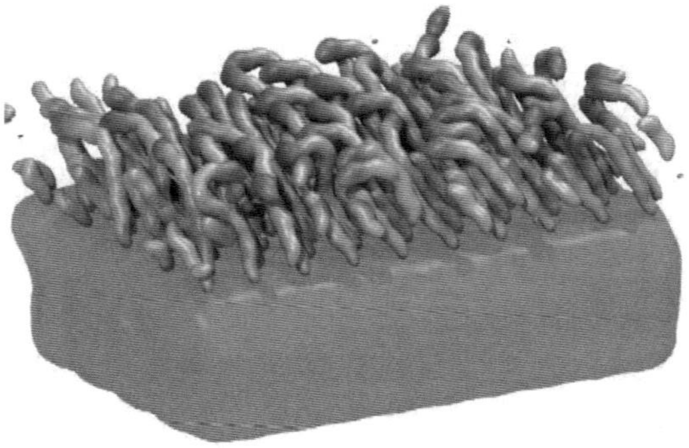

Discussion and Conclusions

Discussion

Good Surfaces

- High water displacement
- Favorable interactions with water molecules
  - High Surface Wetting
  - High QLL Density
- High mobility
  - High Displacement of water from ice to QLL
  - High Water Mobility
- C12C7
- A12A7
- M12H7

Bad Surfaces

- More favorable to interact with themselves
  - Templating
- Lower mobility
- M12C7
- E12C7

Future Work

- Mobility between phases
- Mixing modifiers
  - C12A7
  - C12H7
- Building a Polymer Surface
  - Branching
- Cubic ice simulations
- Passive ice formation
  - Forward flux sampling
  - Nucleation Rate
  - Which surface is best?

Thanks

- Dr. Hadley
- NASA Langley Research Center
  - NASA Grant: NNX12AR25A
- Pennsylvania State University
- GNU Fortran Compiler

Questions?

http://www.sylvanmigdal.com/c/alt/20071231.png

434297 Wetting and Interfacial Properties of CO2/Brine/Illite System

Abstract:

Clays are important because of their association with oil and natural gas reservoirs. Knowledge of clay structure, porosity, interfacial tension and wettability has significant importance in the petroleum industry for better subsurface imaging and fluid flow modeling. Within gas shale, illite comprises up to 20% of the clay minerals, which themselves constitute up to 50% of the overall shale contents. Theoretically, there is an enormous potential for interactions between the clay surfaces and their environments, which makes shale a promising candidate for CO_2 transport and storage.

Wettability is a long-standing issue primarily addressed by looking at the angle of contact at the edge of the interface between a liquid and a solid. The value of the solid-liquid contact angle provides useful information about the interactions between the liquid and the solid. Due to the small particle size and the high humidity sensitivity, the contact angles of aluminosilicate clays are usually difficult to measure. Atomistic simulation provides valuable guidelines for their power to deal with the complexity of large assemblies of interacting components, and therefore becomes a practical approach for the prediction of clay wettability.

The present work focuses on determining the solid-liquid contact angles of fluid droplets on the illite surface using molecular dynamics simulation. In the simulation, the illite surface was built based on the structure obtained from the previous density functional theory (DFT) study. The clay-clay, clay-fluid and fluid-fluid interactions were described by the ClayFF empirical force field. Canonical (constant-NVT) molecular dynamics was utilized to obtain the equilibrium fluid density profiles and the solid-liquid contact angles. Two systems, 1) brine droplet on illite and 2) supercritical CO2 droplet on illite immersed in brine, were studied. The interfacial structure, the fluid concentration profiles, and the liquid-solid contact angles, as a function of brine concentration will be presented.

425695 Reactive Atomic Simulations of Kinetic Friction and Its Velocity Dependence at SiC/SiC Interfaces

Abstract:

Silicon Carbide, SiC, is a promising candidate for advanced braking systems due to its many unique properties such as high mechanical strength, thermal shock resistance, high thermal conductivity, and resistance to both acidic and alkali environments. The efficient design of such brakes for specific applications has, however, been a challenge. This stems from the fact that it is of critical importance to understand how mechanical deformations, relating to friction at the interface between two surfaces, behave. Concurrently, such reactions as oxidation, that also occur at the interface, represent another aspect of complexity to the problem. Such factors and their impact on high-performance brakes are not understood yet. Therefore, without a fundamental, atomic-scale understanding of what happens during the friction process, it would be impractical to optimize the design of SiC brakes for use under extreme conditions. The purpose of this work is to investigate kinetic friction during dry sliding along the atomistic scale boundary between two SiC/SiC bilayers in both a vacuum and in the presence of air.

We have developed a reactive force field (ReaxFF) that describes the interactions between the constituent atoms, the effect of high temperature, and the reactions between the SiC atoms and the air. The amorphous SiC structure was generated during reactive molecular dynamics using ReaxFF. Dynamic simulations have been carried out on a pair of SiC unit cells representing the opposing surfaces at the optimal distance, 2.4 Å, wherein the potential energy is at a minimum. Important properties, such as the energy distribution, the kinetic frictional force, the shearing distance, and shearing deformation, have been calculated as a function of sliding distance at the interface and compared under a variety of conditions.

The interfacial properties were found to play a crucial role in the sliding mechanism. The passivated atoms at the interface, either oxidized or hydrogenated, manifest different thermo-mechanical properties as compared to unsaturated surfaces. Therefore, controlling the surface properties with the information extracted from atomistic-level simulations can greatly help to improve the properties of SiC- type brakes. In addition, thermal effects that result from varying the interface velocity and its subsequent impact on the friction properties of the material were studied.

432010 Interfacial Properties of Model Systems Using Isothermal Isobaric Monte Carlo Simulations

Abstract:

The behavior of fluids at solid surfaces plays a significant role within numerous natural phenomena and industrial applications. To this end, understanding and predicting the interfacial properties of these systems is crucial for the development of emerging technologies. Within our group, we have developed an interface potential based approach for determining interfacial properties. Within our previous efforts, this approach has been implemented within the grand canonical ensemble. However, there are limitations to working within this ensemble, particularly when applying the techniques to complex molecules and working at relatively low temperature. In this presentation, we introduce a means to overcome these limitations. We use a free energy based approach in which Monte Carlo simulations are performed to obtain interface potential by growing a liquid or vapor film on the solid substrate. Isothermal Isobaric ensemble (NPT) Monte Carlo simulation is employed within a "spreading" and "drying" framework to calculate the wetting properties, such as the spreading coefficient, contact angle, and, interfacial tension. Expanded Ensemble techniques are used to evaluate these interfacial properties over a range of temperatures and substrate strengths. We also show how to reduce computational expense of volume moves in NPT Monte Carlo simulation. Results are presented for a model system consisting of fluid-fluid interactions described by Lennard-Jones potential. We also show results for the SPC/E water model at a structureless 9-3 surface over a range of substrate strengths and temperatures. The results obtained for these model systems using NPT Monte Carlo method are compared with results obtained from GCMC simulations.

434118 Characterizing the Curvature of Liquid-Vapor Interfaces and Its Effect on the Kelvin Equation Using Mean-Field Lattice Density Functional Theory

Abstract:

Characterizing the Curvature of Liquid-Vapor Interfaces and its Effect on the Kelvin Equation using Mean-field Lattice Density Functional Theory

Aaron J. Harrison, Stephen P. Beaudoin, David S. Corti

The Kelvin equation describes the relationship between the system saturation and the curvature of the liquid-vapor interface. It is extensively used to describe nucleation in supersaturated systems [1] , and capillarity in unsaturated systems [2]. However, when describing the curvature of a meniscus between two adhering surfaces, the Kelvin equation is an approximation [3–5] governed by the assumption of a pure-component, incompressible system with bulk properties. Hence, it is unclear how well the Kelvin equation describes the capillary forces between two surfaces, particularly when these surfaces are separated by nanoscale distances.

Using a two-dimensional lattice-gas model and mean-field density functional theory, the effect of meniscus curvature between hydrophilic surfaces on the prediction of the Kelvin equation has been studied. First, the dependence of the surface tension on the curvature of the liquid-vapor interface is established for critical bubbles forming within a bulk liquid. It is demonstrated that for a pure-component, bulk system the Kelvin equation properly describes the curvature of the interface at the Gibbs surface of tension, even for very small bubbles. Next, the system is modified to include parallel, hydrophilic surfaces in between which capillary bridges can form. The curvature of these capillary bridges are quantified at differing saturation levels and compared to the Kelvin equation. The deviation from the Kelvin equation as the system approaches zero saturation is quantified as a function of the degree of hydrophilicity of the surfaces and the curvature of the interface. For these capillary bridges, it is found that the radius of curvature is not constant (i.e., the meniscus is not circular) and that agreement with the Kelvin equation decreases significantly as the system approaches zero saturation. Therefore, the Kelvin equation best describes curvatures for pure-component systems or for capillary bridges that are near or at saturation.

[1] A. W. Adamson and A. P. Gast, *Physical Chemistry of Surfaces*, 6th ed. New York: John Wiley & Sons, Inc., 1997.

[2] J. N. Israelachvili, *Intermolecular and Surface Forces, Third Edition*, 3rd ed. Boston: Academic Press, 2011.

[3] J. Melrose, "Model calculations for capillary condensation," *AIChE J.*, vol. 149, no. 1933, pp. 986–994, 1966.

[4] S. Ono and S. Kondo, "Molecular Theory of Surface Tension in Liquids," in *Structure of Liquids / Struktur der Flüssigkeiten SE - 2*, vol. 3 / 10, Springer Berlin Heidelberg, 1960, pp. 134–280.

[5] A. Shapiro and E. Stenby, "Kelvin equation for a non-ideal multicomponent mixture," *Fluid Phase Equilib.*, vol. 134, pp. 87–101, 1997.

425244 Molecular Dynamics Study of the Degradation of Alkylsilane Monolayers Under Shear

Abstract:

Chemisorbed alkylsilane monolayer coatings have been shown to reduce the friction, adhesion, and wear of surfaces under shearing contact; however, long-term lubrication is hindered by film degradation. This degradation is thought to proceed through the rupture of interfacial bonds between individual chains and the substrate and the removal of these chains from the monolayer [1]. As such, conditions which alter monolayer structure, such as normal load, surface roughness, and chain length are likely to influence monolayer degradation. Molecular dynamics (MD) simulations provide a very useful tool for examining how these conditions influence tribological behavior at the molecular level. However, monolayer wear typically occurs on the order of hours [2], which is inaccessible by MD, and typical classical forcefields do not allow for dynamic bond breakage/formation. One strategy to address these issues using classical simulations, is to model monolayer degradation through the random removal of chains and use the reduced monolayer density to sample various stages of the wear process [3,4]. However, Booth et al. [2] have shown that film stability is promoted by increasing monolayer chain length and proposed that the interchain dispersion interactions facilitate longevity; as such, the presence of these broken chains may in itself influence stability.

Here, we use classical MD simulations to examine the degradation of monolayers by randomly eliminating interfacial bonds (i.e., breaking the chains). This allows unbound alkylsilane chains to remain in the system and their mobility to be used as a metric to study wear behavior. Furthermore, previous MD simulations [5] have revealed a twofold increase in friction coefficient for systems that feature realistic surface roughness, suggesting substrate features play a key role in monolayer wear. To explore this, we examine monolayers with defects on both crystalline and amorphous silica at various representative surface roughnesses. In agreement with Booth, et al., we find increasing chain length reduces free-chain mobility, promoting monolayer stability through a reduction in chain-chain energy. We observe enhanced free-chain mobility for monolayers attached to rough, amorphous substrates and also find irregular chain arrangements promote mobility by reducing nematic order. Additionally, we find negligible normal load effects on the wear behavior of amorphous systems as compared to crystalline systems, where shear-induced order is found to promote monolayer stability.

[1] B. D. Booth, S. G. Vilt, C. McCabe, and G. K. Jennings, "Tribology of Monolayer Films: Comparison between n-alkanethiols on Gold and n-alkyl Trichlorosilanes on Silicon.," Langmuir, vol. 25, no. 17, pp. 9995–10001, Sep. 2009.

[2] B. D. Booth, S. G. Vilt, J. Ben Lewis, J. L. Rivera, E. A. Buehler, C. McCabe, and G. K. Jennings, "Tribological Durability of Silane monolayers on Silicon.," Langmuir, vol. 27, no. 10, pp. 5909–17, May 2011.

[3] P. T. Mikulski and J. A. Harrison, "Packing-Density Effects on the Friction of n-Alkane Monolayers," J. Am. Chem. Soc., vol. 123, no. 30, pp. 6873–6881, Jun. 2001.

[4] M. Chandross, E. B. Webb, M. J. Stevens, G. S. Grest, and S. H. Garofalini, "Systematic Study of the Effect of Disorder on Nanotribology of Self-Assembled Monolayers," *Phys. Rev. Lett.*, vol. 93, no. 16, p. 166103, Oct. 2004.

[5] J. E. Black, C. R. Iacovella, P. T. Cummings, and C. McCabe, "Molecular Dynamics Study of Alkylsilane Monolayers on Realistic Amorphous Silica Surfaces," *Langmuir*, vol. 31, pp. 3086–3093, Feb. 2015.

431039 Thermodynamics of Cassie-Wenzel Transitions on Nanotextured Surfaces

Abstract:

In hydrophobic environments, water can become metastable with respect to its vapor; the resulting dewetting transitions influence numerous phenomena ranging from bubble coalescence and colloidal assembly to surface nanobubbles and superhydrophobicity. On superhydrophobic surfaces, the surface texture provides such an environment, and water can either wet the texture in the Wenzel state or sit atop an air cushion in the Cassie state. It is only in the Cassie state that the surface is superhydrophobic, with properties like water-repellency, self-cleaning, interfacial slip, and fouling resistance. However, the Cassie state is often a kinetically trapped metastable state and at elevated pressures, water can descend into the surface texture yielding the non-functional Wenzel state; superhydrophobicity is thus fragile. A promising strategy for improving the robustness of the superhydrophobic Cassie state is to reduce the texture dimensions. As texture size shrinks to the nanoscale, water density fluctuations can become increasingly important. Using molecular simulations in conjunction with enhanced sampling methods, here we investigate how fluctuations affect the free energetics of Cassie-Wenzel transitions on nanotextured surfaces. Our results suggest that highly co-operative water density fluctuations play an important role in determining the thermodynamics, kinetics, and thereby the phase behavior of Cassie-Wenzel transitions on nanotextured surfaces. Our results also allow us to propose improved surface morphologies, which not only destabilize the Wenzel state, but also render it unstable under ambient conditions, thereby facilitating the reversible recovery of the superhydrophobic Cassie state.

426318 An Integrated Theoretical and Experimental Approach to the Design of Liquid Crystal-Based Chemical Sensors

Abstract:

Surface-anchored nematic liquid crystals can undergo orientational changes in response to select chemical stimuli, presenting opportunities for the design of highly sensitive chemical sensors. We present an approach to materials design that combines theoretical predictions, synthesis, and experimental characterization and evaluation for the accelerated discovery of new sensor materials. Our previous work has shown that the surface-induced ordering of the liquid crystal 4-cyano-4'-pentylbiphenyl (5CB) depends strongly on the chemical functionality of the surface, and that an orientational change can be triggered by the presence of small molecules such as dimethyl methylphosphonate (DMMP).[1-3]We developed a computational model for this process, based on quantum mechanics, showing that orientational changes can be predicted on the basis of the relative binding energies of the 5CB and DMMP to the metal cation-functionalized surfaces.

In this presentation, we will describe the use of our computational model to design new classes of liquid crystal-based sensor materials for detection of a wide range of small molecules, by varying the chemical functionalities of liquid crystals as well as the metal cations used in the metal surfaces. We present a number of promising candidate sensors for detecting NO_2, acetone or ClO_2on the basis of their relative binding energies to metal cations. Motivated by our calculations, we have successfully synthesized new liquid crystals with desired chemical functionalities, and experimental evaluation has confirmed the validity of several predictions for enhanced sensor materials. Additionally, we have used thermogravimetric analysis (TGA) and temperature-programmed desorption (TPD) to directly measure the strength of binding of mesogens to metal ions, and used these measurements to evaluate computational predictions.

1. Shah, R. R.; Abbott, N. L., Science **2001**, 293, 1296.
2. Hunter, J.T.; Abbott, N.L., *Applied Materials and Interfaces*, **2013**, 6, 2362
3. Yang, K. L.; Cadwell, K.; Abbott, N. L., Journal of Physical Chemistry B, **2004**, 108, 20180.

431230 Computational Design of MOF Arrays for Gas Sensing: Influence of Array Size on Sensor Performance

Abstract:

Gas sensors are used in applications ranging from assessing the quality of food products to environmental monitoring.[1] Existing sensors typically measure gas composition by indirectly detecting a change of mass or electrical resistance in a polymer or metal oxide substrate upon exposure to a gas mixture. When put in arrays, these gas sensing devices are called "electronic noses" and have improved capability in distinguishing varied gas mixtures. However, both polymer and metal oxide substrates typically have small surface areas, and so the concentration of adsorbed gases is low, which in turn leads to poor gas detection limits. Hence, the record high surface areas, reproducibility, tunability, and overall chemistry of MOFs make them ideal candidates as substrates for improved gas sensors. Furthermore, *arrays* of MOFs can significantly improve sensing capabilities over individual MOF sensors, but so far this strategy has not been systematically investigated. In this work, we show that a wide range of CO_2, N_2, O_2, and CH_4 gas mixtures can be distinguished by combining sensing input from arrays of MOFs of different types. Importantly, we show that although sensing performance is incrementally improved by using multiple similar (or identical) MOFs, dramatic improvements can be achieved by combining MOFs with different structural properties. We assume only a mass based response for each individual sensor, modeled after typical quartz crystal microbalance (QCM) devices, for which we assumed a 5 ng measurement error per sensor. Our findings are that using multiple types of MOFs can significantly reduce the size of the array needed to achieve the same sensing performance as an array of identical MOFs. We analyze the relationship between each MOF combination and how it distinguishes between gas compositions to determine the most optimal array for any given mixture. Finally, we briefly discuss plans towards an experimental implementation of our sensor design.

1. K. Arshak, E. Moore, G.M. Lyons, J. Harris & S. Clifford. A review of gas sensors employed in electronic nose applications. *Sens. Rev.* **24,** 181–198 (2004).

435413 Molecular Simulation of Proteins at Interfaces

Abstract:

Protein adsorption plays an important role not only in a wide range of basic biological processes but also in many applications such as protein chromatography, drug delivery on solid substrates, biosensors, biofuel cells and biomaterials. For these processes and applications, one key issue is the orientation of adsorbed proteins on surfaces. Controlled antibody adsorption orientation on surfaces is necessary to ensure that their active sites are away from the surface and accessible to bulk solution for immunoassay applications. For biofuel cell applications, to enable fast ET, adsorbed cytochrome c should have an orientation with the heme ring close and perpendicular to surfaces. Another key issue that determines the activity of adsorbed proteins is their conformation (i.e., how the conformation of the adsorbed protein resembles that of its native state). In this talk, the protein orientation and conformation on charged surfaces are investigated by a hierarchical approach, i.e., studied by colloidal, coarse-grained and all-atom models. Parallel tempering Monte Carlo and molecular dynamics simulations are used. Effects of surface charge density and sign, and solution ionic strength are examined in our simulations. Simulation results show that van der Waals and electrostatic interactions codetermine the orientation of adsorbed proteins. It is found that the electric dipole and hydrophobic dipole of adsorbed proteins play important roles in determining the protein orientation on charged and hydrophobic surfaces.

433865 Partitioning of Organics into Surfactant Bilayers

Abstract:

The ability of surfactants to self-assemble and structure liquids is frequently leveraged in the design of many products, such as those for personal care and household cleaning. In order for these products to be effective, the influence of the hydrophobic molecules which partition into these micellular structures (e.g., perfumes) needs to be considered. For example, solute loading can lead to a loss in the desired microstructure and subsequently, the macroscopic properties of a surfactant solution. In this work, configurational-bias Monte Carlo simulations in the Gibbs ensemble employing coarse- grained and united-atom models are used to study the partitioning of small organic molecules into a C10E3/water bilayer system. The solubility, spatial distribution, and impact of the small organic molecules on the structure of the bilayer system are discussed.

433081 Adsorption and Electron Transfer of Bacterial Decaheme Cytochrome Mtrf on Gold Surface Studied with Atomistic Molecular Dynamics Simulation, Free Energy Computation and Kinetic Monte Carlo Simulation

Abstract:

The study of electron transfer (ET) of the decaheme cytoheme (MtrF) from the metal-respiring bacteria, *Shewanella oneidensis* MR-1, on substrate surface has important applications for bio-technologies and bio-energy. The orientations and structures of the adsorbed protein on the substrate surface play a key role in ET transfer efficiency. In this work, a hybrid simulation scheme of atomistic molecular dynamics and free energy computations was developed to predict the orientations of MtrF on Au (111) surface. Our simulations demonstrate that protein's adsorption kinetics and orientation are regulated by the fine balance between protein-surface interactions and dehydration of both protein and gold surfaces. The dehydration of gold surface serves as the crucial driving force for the adsorption. The orientation with the heme ring close and perpendicular to the surfaces is one of the most possible configurations. We also predicted ET transfer with Kinetic Monte Carlo simulation by using the most likely equilibrated structures.

431473 Nanoparticle Interactions with Lipid Bilayer Studied with Ghost Tweezers Method

Abstract:

Using dissipative particle dynamics (DPD) simulations, we explore the interactions of nanoparticles with freestanding lipid bilayers. In order to study nanoparticle-bilayer interactions, we introduce a novel "ghost tweezers" technique that emulates a lab experiment performed with optical tweezers. Ghost tweezers represent a virtual harmonic potential applied between the nanoparticle and a chosen system point. The nanoparticle is tethered by the ghost tweezers to a given point by a spring and experiences thermal fluctuations around this point. The average spring force is used to measure the effective force of nanoparticle-bilayer interaction as a function of the particle coordinate. The free energy landscape of the nanoparticle near the bilayer is evaluated as the mechanical work needed to bring the particle from the bulk solution to a particular distance from the substrate surface.

Hydrophobic particles adsorb lipids in a a form of a self-assembled monolayer with the tail facing the hydrophobic particle and heads facing the surrounding water. As the nanoparticle approaches the bilayer, it caused deformation of the latter, associated with a free energy penalty. The deformation is followed by a spontaneous fusion of the freestanding bilayer and the adsorbed monolayer and particle incorporation inside the bilayer hydrophobic segement. Particle escape from the bilayer membrane is also associated with bilayer deformation and a free energy penalty. Using the ghost tweezers method, we explore the free energy landscape of the nanoparticle in the bilayer vicinity and evaluate the free energy barriers associated with particle adhesion and penetration. We vary nanoparticle diameter, degree of hydrophobicity and the surface tension of the bilayer membrane.

The University of Mississippi

Graphene Oxide Nanocarrier for Doxorubicin Anticancer Drug, a Molecular Dynamics Simulation Study

Mina Mahdavi, *Farzin Rahmani, Sasan Nouranian*

15AIChE
Annual Meeting, Salt Lake City, UT

November, 12th 2015

Introduction

The University of Mississippi

- Doxorubicin (DOX), is one of the most effective drugs for:
  - Breast Cancer
  - Bladder Cancer
  - Lung Cancer
  - ...

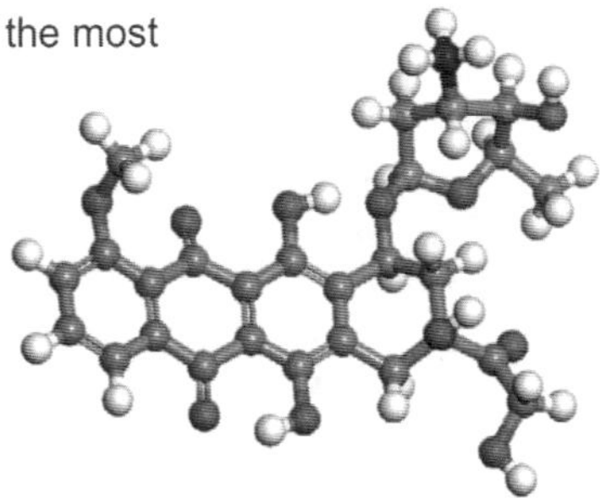

Structure of DOX ($C_{27}H_{29}NO_{11}$)

- Different forms of drug devices have been studied before:
  - Polymeric micelles
  - Dendrimers
  - Liposomes
  - Nanoparticles

1- Xiaoying Yang; Xiaoyan Zhang; Zunfeng Liu; Yanfeng Ma; Yi Huang; Yongsheng Chen; *J. Phys. Chem. C* **2008**, 112, 17554-17558.

Introduction

The University of Mississippi

Graphene (Gr) as a nanocarrier:

- A single-atom-thick sheet
- Is able to adsorb drug molecules on its edges and two large surfaces
- Higher loading capacity of rather than other nanocarriers

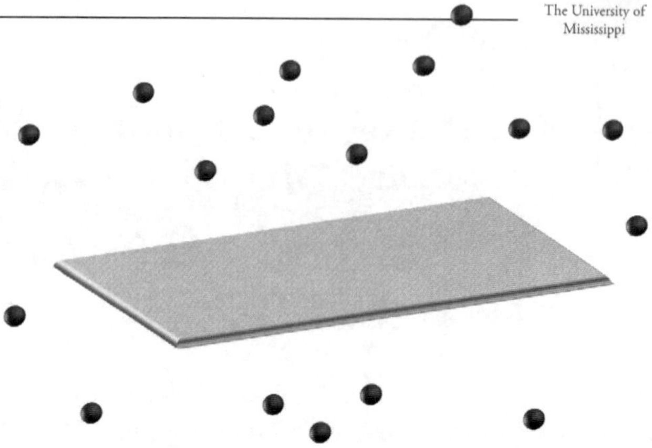

3

Introduction

The University of Mississippi

Graphene Oxide (GO) as a nanocarrier:

Hydrophilic groups of GO,

- Improve the water dispersibility
- Provide a site for covalent modification
- Are able to form hydrogen bonds with drugs

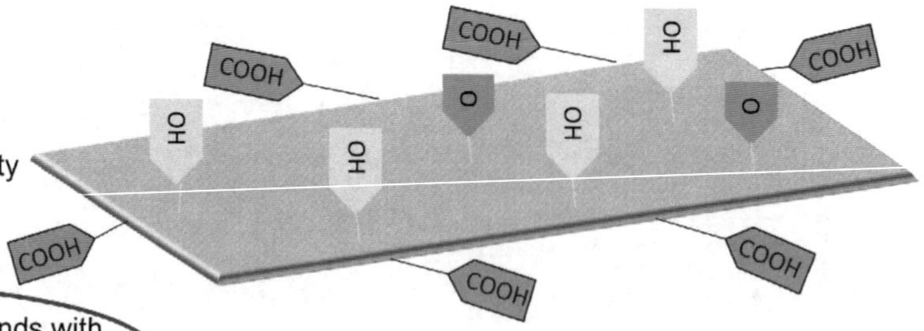

4

Objectives of Research

The University of Mississippi

1. A detailed analysis of **the energetics of the drug/GO interactions** to understand the process of DOX adsorption on the GO in an aqueous solution with **different pH levels**
2. Investigation of **the effects of GO functional groups** on the adsorption characteristics of DOX
3. Investigation of diffusion behavior of drug molecules on nanocarriers

5

Computational Method

The University of Mississippi

- Materials:
  - Doxorubicin :DOX
  - Water
  - Nanocarrier:
    - Graphene : Gr
    - Graphene Oxide (Oxygen/Carbon = 1/6) : GO6
    - Graphene Oxide (Oxygen/Carbon = 1/3) : GO3

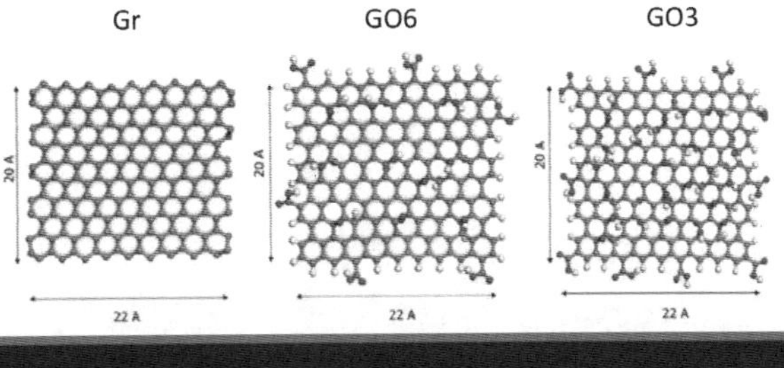

- Simulation cell size: $40\times 40 \times 40A^3$
- Temperature: 310 K
- Force Field : COMPASS II
- Time Step: 1 fs
- Ensemble: NVT
- Number of molecules:
  - Water: 2410
  - DOX: 1
  - Nanocarier:1

6

Computational Method

The University of Mississippi

Calculating the fraction of charged sites on the GO or DOX as a function of pH [1]:

$$N_{charged} = \frac{10^{pKa-pH}}{1+10^{pKa-pH}} \times N_{total}$$

pH	DOX	GO6	GO3
1	Protonated	Normal Structure	Normal Structure
7	Protonated	5 deprotonated groups	10 deprotonated groups
14	Normal Structure	Fully deprotonated	Fully deprotonated

1-A. Adnan, R. Lam, H. Chen, J. Lee, D J. Schaffer,A. S. Barnard, George C. Schatz, D. Ho, W. Kam Liu Mol. Pharmaceutics 2011, 8, 368–374

Interaction Energy

The University of Mississippi

Every 2.5 ps, the interaction energy of GO3/GO6/Gr and DOX is calculated at different pH levels (1,7,14) using the following formula:

$E_{int}= E_{tot}-(E_{GO}+E_{H2O}+E_{DOX})$

E_{int}= Interaction Energy
E_{tot}= Total energy of the system
E_{GO}=Energy of nanocarrier
E_{H2O}=Energy of water
E_{DOX}=Energy of doxorubicin

Interaction Energy

The University of Mississippi

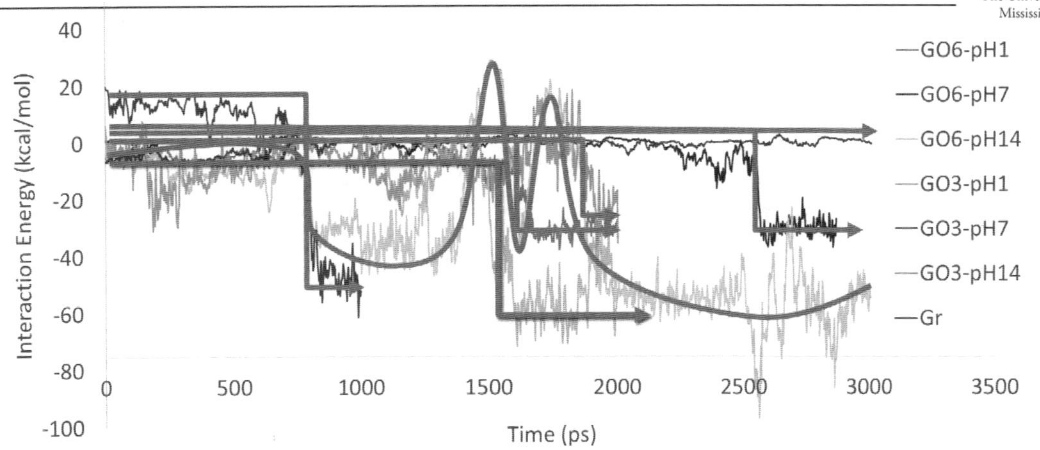

Adsorption site

The University of Mississippi

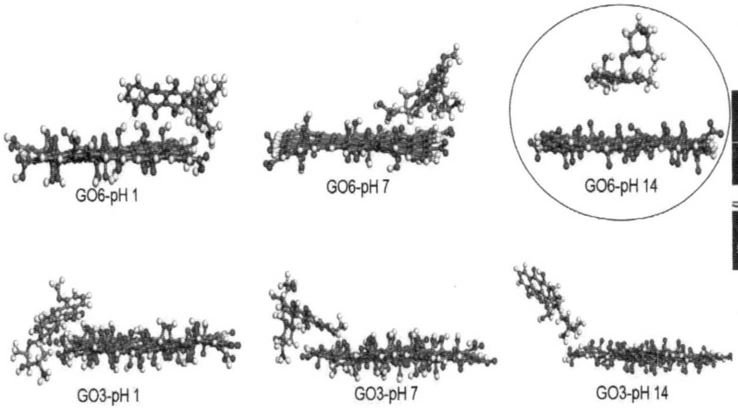

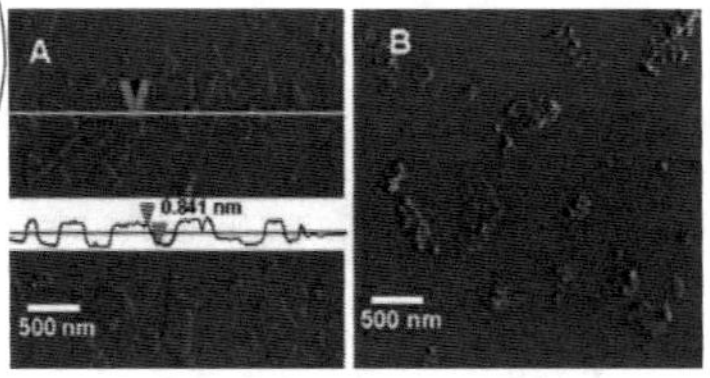

AFM images of GO (A) and GO–DXR (B).[1]

1- Xiaoying Yang; Xiaoyan Zhang; Zunfeng Liu; Yanfeng Ma; Yi Huang; Yongsheng Chen; *J. Phys. Chem. C* **2008,** 112, 17554-17558.

Interaction Energy

The University of Mississippi

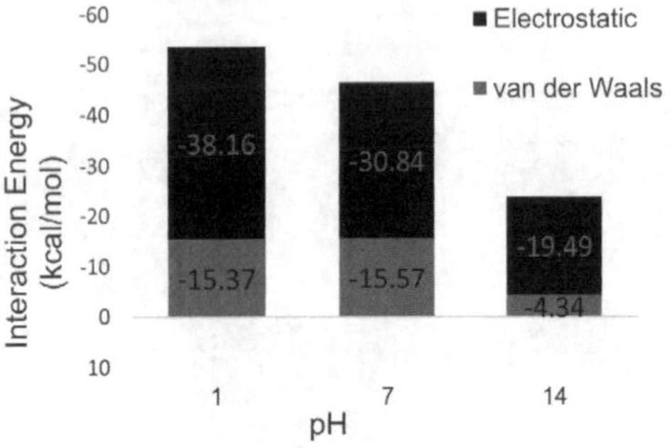

GO3

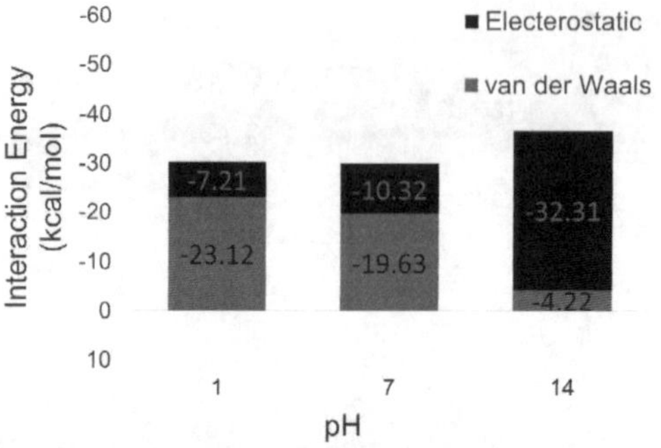

GO6

Diffusion Coefficient

The University of Mississippi

$$D = \lim_{t \to \infty} \frac{1}{6t} < |r_i(t) - r_i(0)|^2 >$$

$r_i(t)$ = Position vector at time t
D = Diffusion coefficient

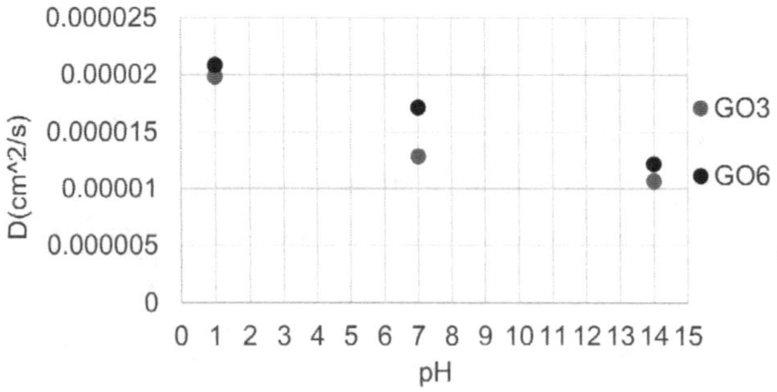

Conclusions

The University of Mississippi

- ➢ Dox molecule adsorbed on GO sheet easier than Gr sheet because of electrostatic interaction.
- ➢ In all case, Dox molecule adsorbed on the edges of GO.
- ➢ Higher pH levels resulted in lower diffusion coafficent.

13

Future Work

The University of Mississippi

- ➢Investigation of desorption mechanism of drug molecules
- ➢Calculation of Hydrogen bonds between drug and nano carriers
- ➢Calculation of loading capacity of GO
- ➢Calculation of size-independent diffusion coefficient

14

The University of Mississippi

Thanks For Your Attention

15

434898 DFT Calculations of Adlayer Structure of Tryptophan on Cu(111)

Abstract:

In this work, we present results of DFT calculations to characterize the structure of moderate to high coverage adlayers

of the amino acid tryptophan on Cu(111) surfaces. This work characterizes the differences in structure between chiral enantiomeric

mixture dosings of the tryptophan on Cu(111). Because tryptophan has a large steric R-group, the 'universally' accepted assumption

of tridentate bonding surface bonding motifs may not be valid in this system; our results help elucidate the truth of this assumption for this (and possibly other related) systems.

430005 On the Requirements for the Formation and Shape of Precursors in Droplet Spreading

Abstract:

We study the requirements for the formation and shape of precursors using molecular dynamics simulations. Using a new method for analyzing our spreading simulations, we identify three regimes with increasing substrate energy: no precursor, a monolayer precursor, and a continuously growing precursor. Changes in free energy of layer formation are reported for the first time and explain the onset and the different types of precursors. Our results also resolve the contradicting claims regarding the role of the spreading coefficient as a requirement for precursor formation.

427772 Stability By Design (SbD): Utilizing Material Science Principles to Predict Physical and Chemical Stability of Pharmaceutical Solids

Friday, November 13, 2015: 8:30 AM
Ballroom D (Salt Palace Convention Center)

Michael Brunsteiner[1], Corinna Gressl[1], Johannes G. Khinast[2], Amrit Paudel[3], Adrian L. Davis[4], Margaret Landis[5], Klimentina Pencheva[4], Garry Scrivens[4] and Geoff Wood[6], (1)RCPE, Graz, Austria, (2)Research Center Pharmaceutical Engineering (RCPE), Graz, Austria, (3)Advanced Products and Delivery, RCPE, Graz, Austria, (4)Pfizer Worldwide R&D, Sandwich, Kent UK, Kent, England, (5)Pfizer Worldwide R&D, Groton, CT, Groton, CT, (6)Pfizer

Stability by Design (SbD): Utilizing material science principles to predict physical and chemical stability of pharmaceutical solids

M. Brunsteiner [1], C. Gressl[1], J.G. Khinast[1], A. Paudel[1]

Adrian L. Davis [2], Meg Landis [3], Klimentina Pencheva [2], Garry Scrivens [2], Geoffrey P.F. Wood [3]

[1]Research Center Pharmaceutical Engineering, Graz, Austria

[2] Pfizer Worldwide R&D, Sandwich, Kent UK

[3]Pfizer Worldwide R&D, Groton, CT, USA

Ensuring, both, physical and chemical long-term stability of the drug product is one of the key requirements in successful drug formulation development. While significant knowledge exists on solution-state stability of pharmaceuticals, the predictive knowledge of solid-state physical and chemical stability in solid oral dosage forms is limited. Here we present two examples of our research towards achieving stability by design (SbD) for solid oral dosage forms through the use of *in-silico* methods to accelerate and streamline formulation development.

As a model for chemical drug degradation we consider Carvedilol (CAR), a compound used in the treatment of congestive heart failure. The compounds secondary alcohol has been shown to undergo esterification when formulated with citric acid (CA)as an excipient.(1) The free base form of CAR can crystallize in at least two different polymorphic forms. Based on structural and energetic results from Molecular Dynamics (MD) simulations of the crystal surfaces of two different CAR polymorphs in contact with vacuum or citric acid, we calculated a range of descriptors, including surface energies and exposed reactive surface areas. A thorough analysis of these descriptors and their relative magnitudes for different crystal surfaces of the CAR polymorphs allows us to investigate the relative importance of degradation of molecules on crystal surfaces versus degradation of molecules that are highly mobile due to surface defects and partial amorphization. Accordingly, two different crystalline forms (included in MD simulation) and an amorphous form of CAR were generated of two different size ranges. Following the comprehensive solid-state and particulate characterization, CAR-CA blends prepared using different solid-states and sizes of CAR were subjected to accelerated condition of 50°C. The comparison of the experimental data on kinetics and extent of CAR-CA interaction products formation with the descriptors derived from MD simulations will provide the predictability of chemical stability by the computed descriptors.

As models for physical drug stability, we consider a series of API-polymer combinations that have been used to produce amorphous solid dispersions (ASD), (2) a commonly used strategy in the formulation of poorly soluble drugs. We performed extensive MD simulations of ASDs at various drug loads and calculated descriptors such as relative hydrogen boding propensities, free volumes, API mobilities, and API-polymer intermolecular interactions. A comparison of the obtained results with existing literature data for the relative stabilities of the different ASDs shows that both kinetic and energetic factors can play a role in determining the stability of ASDs. We suggest criteria that can be determined *in-silico* and used to estimate the relevance of these factors for a given API-polymer combination.

References

1. Larsen, J., Cornett, C., Jaroszewski, J. W., and Hansen, S. H. (2009) Reaction Between Drug Substances and Pharmaceutical Excipients: Formation of Citric Acid Esters and Amides of Carvedilol in the Solid State., Journal of Pharmaceutical and Biomedical Analysis 49, 11–17.

2. Eerdenbrugh, B. V., and Taylor, L. (2010) Small Scale Screening to Determine the Ability of Different Polymers to Inhibit Drug Crystallization upon Rapid Solvent Evaporation, Molecular Pharmaceutics 7, 1328–1337.

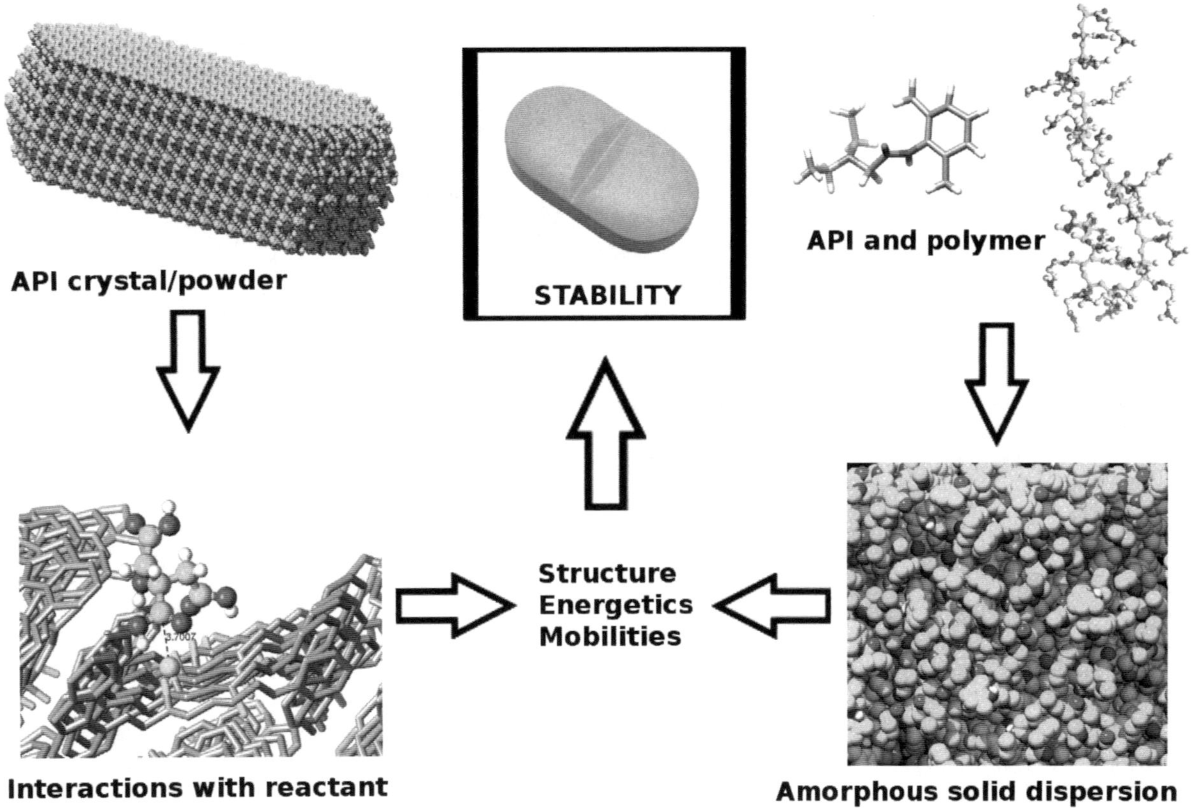
API crystal/powder
STABILITY
API and polymer
Interactions with reactant
Structure
Energetics
Mobilities
Amorphous solid dispersion

425319 Exploring Polymorph Free Energy Landscapes with Hamiltonian Reweighted Molecular Dynamics

Abstract:

The presence of multiple stable crystal structures in solid organic materials can limit their commercial viability when two or more observable polymorphs exhibit markedly different physical properties. Crystallography researchers estimate that 50% of all drug molecules exhibit polymorphism, but only 1% of the materials with a well-characterized crystal structure in national crystal structure databases contain a similar level of data for an alternative polymorph structure.

Current computational methods for predicting polymorphic behavior generate many possible crystal structures and rapidly evaluate the lattice energies of these candidate structures to determine the most likely to be observed experimentally. However, lattice energies alone do not fully capture the stability of systems that are at room temperature and sample a range of configurations due to thermal motion. In addition, *ab initio* quantum mechanical models or at minimum polarizable force fields are likely necessary to represent the physical system, but the extensive sampling required in molecular dynamics can currently only be performed with classical point-charge force fields.

In this work, we present an atomistic simulation approach to evaluate the free energy difference between crystal polymorphs in both computationally cheap and expensive Hamiltonians. Molecular dynamics simulations are conducted in a classical point-charge all-atom potential designed to overlap with the expensive Hamiltonian. Multistate Hamiltonian reweighting is then employed on these configurations post simulation to estimate the free energy difference in more expensive force fields without any additional sampling. We therefore access computationally demanding free energy differences by avoiding direct simulation within the polarizable potentials.

This simulation technique is applied to solid phase benzene, which is known to have multiple distinct polymorphs. We reweight to polarizable potentials and the simulation results indicate that the exact polymorph stability is strongly sensitive to the choice of Hamiltonian. Furthermore, the successful Hamiltonian reweighting technique demonstrates that the relevant phase space of polarizable Hamiltonians in this system can be approximated using only point-charge models. These preliminary results represent a significant step toward the ultimate goal of tractable thermodynamic calculations using more expensive many-body energies.

431785 Molecular Modeling to Design and Control (M2DC): A First Principles Approach to Polymorph Prediction and Crystallization Unit Design

Abstract:

Crystallization remains one of the most important industrial separation processes to date. Despite its wide spread use, a fully fundamental understanding of the nucleation process is lacking. The classical nucleation theory (CNT) rate expression, formulated by Gibbs, Weber, and Volmer, at the beginning of the 20^{th}century, remains the main framework against which experimental evidence is compared, despite being overly simplistic, and failing to capture data well in a range of systems. Skepticism towards the degree to which thermodynamic bulk properties used in the CNT rate expression can be even applied to nano scale clusters is warranted. Rather, with the advent of modern supercomputers, the possibility to relax the suit of CNT assumptions in favor of fully atomistic calculations provided via molecular dynamics is becoming a more and more appealing alternative. From a solid-state physics perspective, polymorphism, the ability of a molecule to exist in multiple crystalline phases, remains a thorn in the side of both industrial crystallization experts, and molecular modelers alike. Landscape energy minimization techniques, although elegant in their own right, fail to include standard design variables such as solvent, temperature, or pressure. Molecular dynamics, which encapsulates the aforementioned design variables naturally, has suffered from time scale challenges associated with activated rate processes, such as nucleation. However, with the advent of modern supercomputers, and HPC hardware such as xeon phi coprocessors, these types of simulations are becoming more and more feasible. The crystallization community would more than like molecular simulations to become primetime, because in the absence of in silico polymorph specific nucleation rates, expensive solvent screening experiments must be performed. Furthermore, the question of "have we truly found all possible forms" is never truly answered by such experiments.

In this work, we investigate the potential of atomistic information provided by hybrid monte carlo molecular dynamics (MCMD) to be used to not only predict polymorphism as a function of operation conditions, but also the degree to which polymorph specific solubility, and metastable zone curves can be calculated. These calculations will serve as inputs into control models, allowing a fully atomistic design of a crystallization unit. As microfluidic devices provide a means to measure homogeneous nucleation rates using minimal reagents and in minimal time, they will serve as a benchmark for determining the degree of accuracy of our homogeneous nucleation rate calculations prior to further scale-up for industrial applications.

412913 PAT for Form Monitoring during Drying

Abstract:

PAT for Form Monitoring During Drying

M. Dimitri, S. Allwein, P. Mudipalli, and M. Christie

Teva Branded Pharmaceuticals, Inc.

Chemical Synthesis Center

383 Phoenixville Pike

Malvern, PA 19355

A.I.Ch.E. Annual Meeting 2015

Session: "PAT for Process Understanding and Elucidation of Fundamental Phenomena in Drug Product/Substance Development"

Abstract: Controlling and monitoring polymorphic form during drying of active pharmaceutical ingredients (APIs) is increasingly an area of interest particularly for specialized drying, such as humidified drying. The use of PAT instruments is often powerful in understanding the drying kinetics of an API, identification of the final polymorphic form in situ in the dryer, and determination of drying end point. This study examines the development and use of Raman Phat Probe technology to monitor a humidified drying process of an API and examines some of the key scale-up parameters involved. For this API, three different known polymorphs are possible during drying, with one polymorph being irreversible once formed. In addition, online analysis was determined to be critical as the API is extremely sensitive to relative humidity and form conversion occurs rapidly, making offline analysis challenging. This study demonstrates the use of raw spectra and chemometric analysis in monitoring and defining the drying kinetics of the two main forms of interest. The example of the use of this technology that is being presented shows its applicability both for drying process development and for confirmation of form at scale (ie, in situ in the dryer).

Keywords: Drying, Phat Raman, Form Monitoring, PAT

443949 Molecular Modeling of Additive Effects on Calcium Pyrophosphate Crystals

Abstract:

Calcium pyrophosphate (CaPP) deposition disease is a rheumatologic condition caused by the formation of calcium pyrophosphate crystals in the joints. CaPP crystals form in joint cartilage as the results of an unknown metabolic disorder. Eventually, CaPP crystals are ejected from the cartilage into the surrounding synovial fluid where they continue to grow. CaPP deposition disease becomes increasingly common with age, affecting approximately a quarter of people over 85 years old.[1] Despite its prevalence, there are no treatments to inhibit the growth or formation of CaPP crystals. Mechanistic understanding of how additives interact with crystal faces is important for the development of additives to inhibit CaPP crystal growth.

We use atomistic simulations to study the effects of additives on calcium pyrophosphate crystal faces. Accurate modeling of additive effects on CaPP crystal faces requires a molecular force field that can reproduce the experimental crystal structures of CaPP. We develop a new force field for pyrophosphate using the functional form of the CHARMM force field. Starting with parameters in the CHARMM General Force Field, we refit bond, angle, and dihedral parameters to match ab initio simulation data. Finally, the non-bonded parameters were optimized to minimize the root mean squared deviation between the experimental and force field derived crystal structure of the thermodynamically stable $Ca_2P_2O_7$Ÿ$2H_2O$. The resulting force field accurately reproduces the crystal structure of all solved hydrated CaPP forms.

Simple salts have been shown to affect polymorph selection and crystallization of CaPP. We study the effects of Ca^{2+}, Mg^{2+}, and Na^{+} on CaPP, by calculating their binding free energies to CaPP surfaces using adaptive biasing force and free energy perturbation simulations. Ion binding energies are calculated for fast growing CaPP crystal faces for crystals grown homogeneously and heterogeneously on model biological surfaces. Positive ions that bind more strongly to fast growing crystal faces than Ca^{2+} may inhibit crystal growth. We find that Mg^{2+} binds more strongly than Ca^{2+} to the fast growing faces of β-$Ca_2P_2O_7$Ÿ$4H_2O$, while Ca^{2+} binds more strongly than Mg^{2+} to the fast growing faces of the thermodynamically stable $Ca_2P_2O_7$Ÿ$2H_2O$.

[1] Guerne, P. A., and R. Terkeltaub, "Calcium pyrophosphate dihydrate crystal deposition: epidemiology, clinical features, diagnosis, and treatment," Chapter 21 in Terkeltaub, R., ed., "Gout and Other Crystal Deposition Arthropathies", Elsevier Heath Sciences, Philadelphia, PA (2012).

AUTHOR INDEX

AUTHOR INDEX

AUTHOR INDEX

AUTHOR INDEX

AUTHOR INDEX